THE GOAL SHOULD BE, NOT TO IMPLANT IN THE STUDENTS' MIND EVERY FACT THAT THE TEACHER KNOWS NOW; BUT RATHER TO IMPLANT A WAY OF THINKING THAT ENABLES THE STUDENT, IN THE FUTURE, TO LEARN IN ONE YEAR WHAT THE TEACHER LEARNED IN TWO YEARS. ONLY IN THAT WAY CAN WE CONTINUE TO ADVANCE FROM ONE GENERATION TO THE NEXT.

EDWIN JAYNES

I AM, SOMEHOW, LESS INTERESTED IN THE WEIGHT AND CONVOLUTIONS OF EINSTEIN'S BRAIN THAN IN THE NEAR CERTAINTY THAT PEOPLE OF EQUAL TALENT HAVE LIVED AND DIED IN COTTON FIELDS AND SWEATSHOPS.

STEPHEN JAY GOULD

JAKOB SCHWICHTENBERG

TEACH YOURSELF PHYSICS

NO-NONSENSE BOOKS

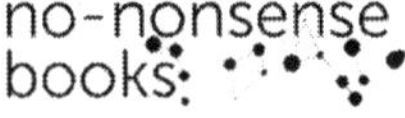

First printing, May 2020

Copyright © 2020 Jakob Schwichtenberg

With illustrations by Corina Wieber

All rights reserved. No part of this publication may be reproduced, stored in, or introduced into a retrieval system, or transmitted in any form or by any means (electronic, mechanical, photocopying, recording, or otherwise) without prior written permission.

UNIQUE ID: 886ae0c2088124be43d800396b71a97bd05d8b03f473c4b11026da036da13595
Each copy of Teach Yourself Physics has a unique ID which helps to prevent illegal sharing.

BOOK EDITION: 1.1

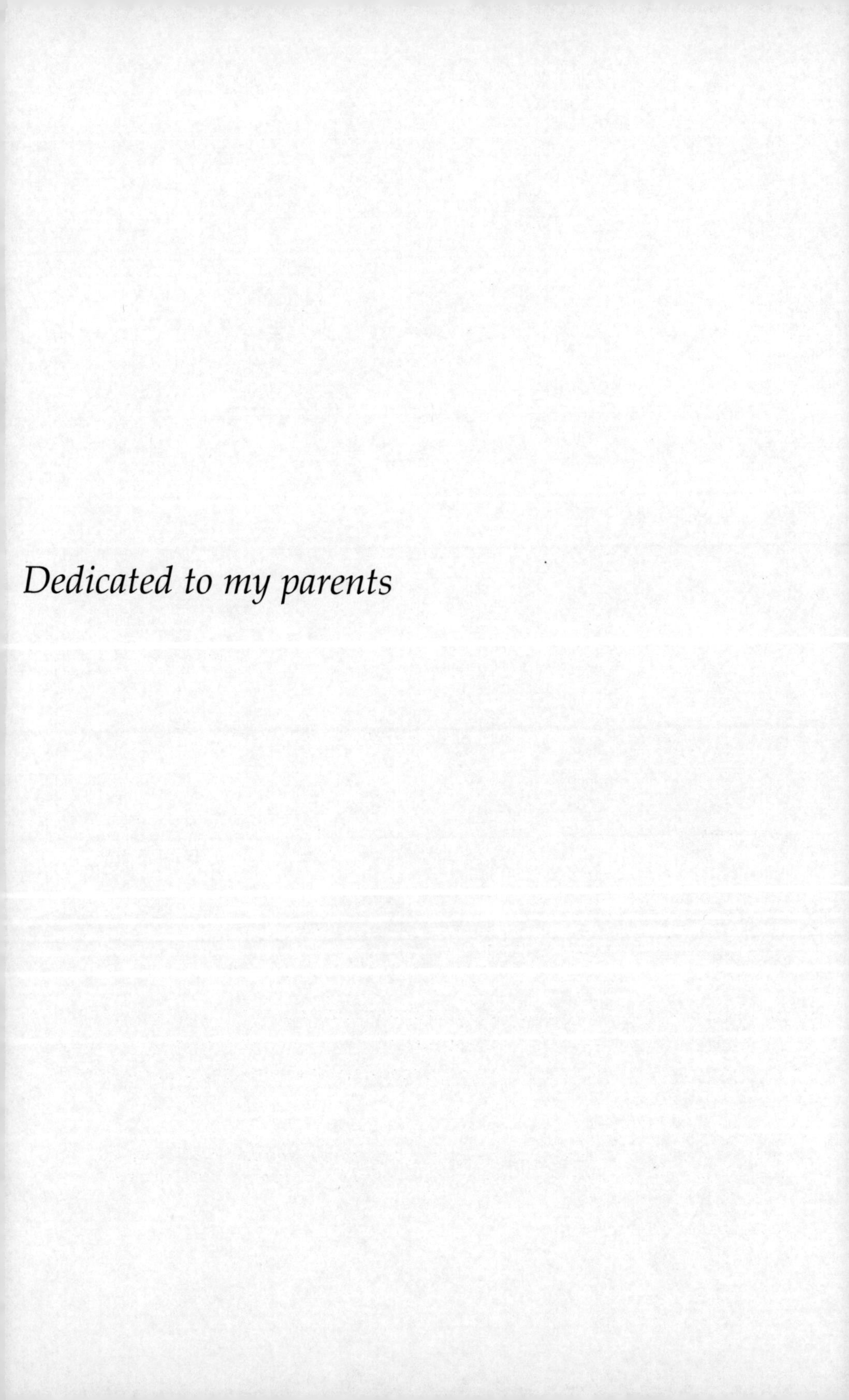

Dedicated to my parents

Preface

This is neither a career guide nor an attempt to explain everything you need to know about physics in a single book. There are many excellent career guides out there and if your goal is to improve your chances of securing a permanent academic position, you should pick one of those.[1] Moreover, after reading this book you will still have to learn more about physics using other resources like textbooks or lectures.

[1] An excellent career guide is:

Peter Feibelman. *A PhD is Not Enough! : A Guide To Survival in Science*. Basic Books, New York, 2011. ISBN 9780465022229

Instead, this book is a handbook containing all the advice and recommendations about learning physics I wished someone had told me when I was younger.

My goal is to encourage more people to think deeply about nature by using the appropriate tools. If a larger number of people understand the fundamentals of physics, our chances are much higher that someone comes up with a good solution to our present obstacles. And I'm convinced that anyone can contribute something meaningful to the world of physics.

Of course, it's equally possible to simply enjoy learning about physics without the explicit goal to contribute something yourself. It doesn't matter whether you're a full-time student of physics or someone who wants to understand more about nature in your spare time. Understanding physics is a goal within everyone's reach.

This book is the result of discovering how few longtime students of physics have developed a truly deep understanding. During the past few years, I've talked to hundreds of students from all corners of the world. What I learned is that, even after many years of studying, many students weren't making the kind of progress they should have. The sincerity of their aspirations and the amount of time they spent on learning were definitely not the problem. Instead, what they lacked was a clear understanding of the most basic objects, of how physics really works, why it works the way it does, and how the various parts of physics are connected.

Put another way, what they needed but didn't have was a clear map.

This book is my modest attempt to improve the situation and I hope it offers a portal anyone can use to get access to the amazing world that we call physics. Think of this book as a traveler's guide, providing you with maps of the territory and detailed directions for getting where you want to go. By using it as your travel companion you will always know where you're at, and where you need to head next. I'm convinced that this will help to make your journey more enjoyable.

With that said, I have to admit that I was quite hesitant to write this book because I felt it was presumptuous of me to tell others what and how to learn. I certainly don't have all the answers.[2] But what finally convinced me was the realization that no such travel companion currently exists

[2] My background will be discussed in a bit more detail in the following chapter.

and I would have loved if a book like this had existed when I started studying physics. I also realized that I don't need to write the ultimate authoritative reference to provide something useful. Instead, I will simply share everything I wished I had known earlier and each reader can make of that whatever he or she wants.

So while this book is certainly not perfect, at least it's a start. In fact, I think it would be fantastic if someone reads this book, disagrees with most of my recommendations and then writes a much better travel companion for physics. Furthermore, I believe that this book is only successful if it sparks new ideas. So please don't take anything in this book at face value and don't blindly follow my advice. Question everything and try to find better alternatives. If enough people do this and share their thoughts, it will become easier and easier for everyone to understand how nature works. After all, the goal is not to spread my ideas but to make it easier for subsequent generations to learn physics. I'm just trying to provide an initial spark.

In any case, I'm glad to have you aboard, fellow voyager! Hopefully, you'll enjoy reading this book as much as I have enjoyed writing it.

Karlsruhe, August 2019 Jakob Schwichtenberg

PS: If you have any kind of feedback, please do not hesitate to contact me. I'm available via email at mail@jakobschwichtenberg.com.

About the Author

I rarely talk about myself because I'm convinced that a subject like quantum field theory, as one example, is a far more interesting topic. This book, however, contains many personal opinions. Therefore, I thought it may be helpful for some readers to know my background.[3]

[3] This section contains no details that are essential for the rest of the book, so feel free to skip it.

Perhaps, you are wondering why you should care about what I have to say about physics. I'm not famous, and I will not win the Nobel Prize in the foreseeable future. So, it's quite reasonable to question why anyone would be interested in my advice.

First of all, I'm convinced that for the content of the book that you're reading right now, it's actually an advantage that I'm neither a genius nor a world-renowned expert.

A genius struggles with different things than an ordinary person and may find it hard to relate to the problems of "normal" people. And, of course, a normal person will struggle to understand the problems a genius encounters. So, if you're a genius, you're probably wasting your time here.[4] But if you're a (relatively) normal person, this book may be exactly what you've been looking for.

[4] We will talk about the role of talent below.

Similarly, experts have often forgotten what it's like to be a beginner. And so, it's probably good for the content of this book that I just finished my Ph.D. and still remember all the problems non-experts regularly encounter.

With that out of the way, let's get a bit more personal.

As you've probably seen on the cover of this book, my name is Jakob. Hi!

I've already revealed that I recently finished my PhD (in theoretical particle physics on "The Strong CP Problem and Non-Supersymmetric Grand Unification", in case you're interested in such details).

During my master's studies, I wrote a textbook called *Physics from Symmetry* which was published by Springer. The book became, in some ways, a surprise bestseller. The biggest contributing factor to its success was, as already discussed above, that I'm neither an expert nor a genius. I was simply a regular student who wrote his thoughts down to help other students understand.

Motivated by the success of this book, I wrote several other student-friendly textbooks during my PhD, which all turned out to be quite popular among readers all around the world.

Rest assured that I'm not writing this to brag. It's just that the books and my PhD are all I can offer in terms of credentials.

But maybe you also want to hear a bit more about my personal background and, in particular, why I felt the urge to write this book.

Max Frisch once remarked,[5]

> "Everyone sooner or later invents his story, which later becomes his life."

[5] Max Frisch. *Gantenbein : a novel*. Harcourt Brace Jovanovich, San Diego, 1982. ISBN 978-0156344074

So, here's mine:

I was quite bad in school. Not so bad that I got into trouble, but still pretty bad. I was bored most of the time and only invested the minimal amount of time necessary to pass the exams. To give an example, in the tenth grade I got a *D* in mathematics. This is exactly the grade you need to pass without getting into trouble.

Just one year later, I got mostly straight A's in mathematics and started studying physics shortly after. To this day, I have spent most of my time thinking about mathematics and physics.

How did this change happen? How did I transform from a bad student into a passionately interested student?

Looking back, I think one of my problems was that I never understood why I should care about the things the teacher talked about. While for many students getting good grades was motivation enough, this never worked for me. In a way, I understood too early that grades don't matter.

I remember asking my teachers several times, especially in mathematics, why we were taught certain topics. "What can we do with this? Why is this important?" I didn't get satisfactory answers and was rewarded with bad oral grades.

Then something huge happened.

At the time I frequently visited flea markets with my parents. And one day when I was sixteen I bought a book titled *Surely You're Joking Mr. Feynman*[6] at such a market. It was the best euro I ever invested. The book turned my world upside down. Before reading it, I was mostly interested in computer games and soccer. But with Richard Feynman's help, I finally understood why people care about mathematics and physics. I started to understand that physics and mathematics aren't the boring things that the teacher presented. Instead, they are fun and important. In physics, you are trying to understand what makes nature tick at the most fundamental level. And mathematics is the language that you need to talk about physics. These were the answers I was craving. I suddenly got interested in all those tricks that the math teacher presented because I was curious about how they can help to understand nature. Suddenly, school was fun.

[6] Richard Feynman. *Surely You're Joking, Mr. Feynman!" : Adventures of a Curious Character*. W.W. Norton, New York, 1985. ISBN 978-0393316049

Around the same time, another huge transformation happened after a soccer match. One Friday, I was asked to help out a team in a much higher league and play in a match where usually only guys at least two years older than me played. The match went okay. I didn't play exceptionally well, but I didn't make any noteworthy mistakes either. Nevertheless, when I played a match on my normal team the next day, I was completely transformed. Everything suddenly seemed so simple. I won every tackle, got three assists and scored two goals myself. And this was in a match against the supposedly best team in our league.

The week before, I was just a regular player with no outstanding abilities. But after this one match with the older guys, I was suddenly a superstar. And the only thing that had changed during that week was how I thought about myself.

Previously, I had thought about myself as an ordinary player and consequently, I played like one. But by competing in the match with the older guys, I realized that I was capable of competing at a much higher level. Through this subtle shift in mindset, my abilities had increased tremendously overnight.

As I analyzed all of this in the weeks that followed, I started to realize that similar transformations might be possible in other areas of my life too. For instance, I had never thought about myself as a particularly smart person. I had always believed that my slightly below average grades proved that I was simply an ordinary student. But what if my cognitive abilities, just like my soccer abilities, were merely limited by how I thought about myself?[7] After realizing this, my grades started to improve dramatically. Moreover, I suddenly started to believe that I was capable of things like becoming a physicist which I had previously discarded as completely unrealistic.

[7] I now know that what I learned through this experience is a well-researched topic. In the terminology introduced by Carol Dweck, I had switched from a fixed mindset to a growth mindset, c.f. [Dweck, 2006]. People with a "fixed" mindset believe that we *are* either great or flawed. People with a "growth" mindset believe our greatness (or flaws) are because of our actions. Our mindset is a result of our environment. When our parents say, "You are great," instead of "You did great work," they accidentally create a "fixed" mindset.

The lessons I learned during that time were reinforced by books like Malcolm Gladwell's Outliers[8].

[8] Malcolm Gladwell. *Outliers : The Story of Success*. Little, Brown and Company, New York, 2008. ISBN 978-0316017930

The main point I took away is that talent does not matter as much as people want you to believe. This has been demonstrated in countless studies now.[9] What is usually stopping us is our self-image and insufficient motivation.

[9] If you still believe in the importance of talent, you should read Peak by Anders Ericsson and Robert Pool [Ericsson and Pool, 2016]. Ericsson has carried out much of the original research on which books like Outliers, Talent is Overrated, The Talent Code, The Practicing Mind and The First 20 Hours are based.

We construct our self-image using all the information about ourselves that we have collected in our lives. The most important source of information is how others react to our behavior. If a reaction comes from a person we value, like a parent, we will pay special attention and update our self-image accordingly.

These reactions are not a very reliable source of information and therefore, self-images are always wrong.[10]

More specifically, we usually don't take into account that there are lots of reasons why people sometimes react weirdly.

For instance, let's imagine a mother who loves her daughter and supports her in any way she can.[11] But one day she comes home after a bad day at work, completely tired and with a headache. Her daughter isn't aware of how her mother is feeling, so she jumps around and sings all the new songs she learned in kindergarten. After a while, her mother loses control for a short moment and screams, "Shut up! Your shrill, ugly voice is so annoying!"

The girl doesn't understand that it's not actually her voice that her mother finds so annoying. Any noise would have had exactly the same effect on her. But since the girl loves her mother, she takes her words seriously. As a result, she stops singing and becomes extremely shy at school because she is afraid that others will find her voice equally horrible. And once she is known as the shy girl in school, she stays in her role as the shy girl all through her life.

That's, of course, just one totally hypothetical example. But it's easy to imagine similar situations that result in people believing they are dumb or ugly. And once you've developed a self-image as a dumb, shy, or ugly person, it's really hard to change it. You will start to act like a dumb, shy, or ugly person and others will react accordingly. This, in turn, confirms your self-image and so on.

While a positive self-image can be a powerful tool, a negative one is like a prison. Moreover, it's difficult to change a once developed self-image since other people act like prison guards.

[10] A self-image is a model we have of ourselves and as the famous aphorism in statistics states: "All models are wrong, but some are useful". The key is to realize that this is also true for your self-image and to use it to your advantage.

[11] The following example is inspired by a section in

Miguel Ruiz. *The Four Agreements : a Practical Guide to Personal Freedom*. Amber-Allen Publishing, San Rafael, California Carlsbad, California, 1997. ISBN 9781878424310

In some sense, we are like those elephants who never try to walk away because they were always stopped as babies using a small rope. Now that they're all grown up, they could easily rip the rope apart. But they don't do this because they still believe they wouldn't succeed.

Our self-image is the rope that's holding us back and it's constructed using notions like talent.

But why then are concepts like talent so popular?

Because they are convenient. On the one hand, they can serve as a convenient excuse for why we aren't able to do certain things. We think,

"Well, I simply have no talent for mathematics", or

"I'm not smart enough."

On the other hand, they are used by "smart" people to feel good about themselves thinking,

"Haha, look at these dumb people."

Most people never reach their potential because they think they don't have what it takes. And many are convinced that physics is one of the most difficult subjects that you can study. Statements like the following are extremely common:

"I'm not smart enough to learn physics."

"I'm not logical enough to be good at it, even if I learned."

"I simply don't 'think' the way physicists think."

"I could never catch up to their skill level if I start so late."

But all of this is nonsense.

People think this way about physics because it's usually presented way more complicated than it really is and others act way smarter than they really are. Be assured that every theory in physics is, at its heart, simple. You don't need any special talent to understand them and it's never too late to start.

Now, after this short tangent, back to my story.

A few years later when I was attending university, I struggled again, even though I was motivated and thought of myself as someone who was capable of becoming a physicist.

Most lectures were just too boring and the professors didn't talk about the things that really interested me. In addition to that, we were forced to spend most of our time doing complicated calculations as homework exercises. And to make matters worse, all the exams only tested our ability to carry out complicated calculations and to apply sophisticated mathematical algorithms. Thus, to pass the exams we needed to practice these algorithms and methods as much as possible. We spent days in the library working our way through the past year's exercises. That's really the only thing we all did during that time. It wasn't fun and even though I ultimately passed the exams with acceptable grades, I became increasingly convinced that I didn't really understand any of the physics I was supposed to learn about.

And I'm certainly not the only one who has experienced

this. Jeff Schmidt, for example, describes physics graduate school as some kind of "bootcamp based on homework"[12].

[12] Jeff Schmidt. *Disciplined Minds*. Rowman & Littlefield, Lanham, Md, 2000. ISBN 9780742516854

Sure, the lectures were called "Classical Mechanics" or "Quantum Mechanics I". But at the end of them, I had primarily learned how to apply specific methods to describe various systems. But where do these methods come from? Why do they work the way they do? These questions were never satisfactory addressed, and there was no room for discussions or questions.

During the lectures, the professors repeatedly fill the blackboards with formulas by copying the handwritten notes they prepared in advance. The students, in turn, are busy writing down each boardful before it gets erased to make room for even more equations. At the beginning of a course the professors usually claim that they're happy to answer any question the students may have. But if a student then really asks a question, the professors' rushed answer clearly signals that this is not quite true. Each question is treated like a disturbance since it takes time from their primary goal: to copy all of their notes onto the blackboard before the semester is over. Thus, the only questions that get asked after the first lecture are about minor points of clarifications like "Shouldn't that be a plus sign?" or "Is that a two in the denominator?"

At this time, however, I believed that the university curriculum followed some grand plan, and we were all just not ready to realize it. I was truly convinced that these seemingly futile exercises and lectures would turn out to be invaluable in the future. If we'd just stick it out and work our way through a few more calculations, the great epiphany would come and everything would fall into place.

After two years, I was so frustrated with my lack of understanding that I decided to try to figure things out on my own. I went to the library and skimmed through whatever quantum mechanics textbook they had available. That didn't work very well. I became even more frustrated because the style and content of the textbooks were basically indistinguishable from the one I had experienced in lectures. No matter which book I picked, they were dry, boring and didn't make much sense.

What finally rescued me once more was a book by Richard Feynman. The third volume of the Feynman Lectures on Physics was exactly what I was looking for. In contrast to all the other books I had read previously, it was not only fun to read but also answered many of the questions that had bothered me for quite a while. And most importantly, Feynman always carefully explained why a given concept was discussed and where it was coming from. Afterwards, I quickly read the other two volumes of the Feynman lectures and understood during those weeks more than I had the previous two years of following the university's curriculum.

Of course, Feynman's lectures were just a starting point. They are quite old, only cover a few specific topics and weren't able to answer all of my questions. But they showed me what is possible and opened the door for me to the wonderful world of self-learning.

After this discovery, I skipped all lectures and spent my days teaching physics to myself. Finally, learning physics was fun, and I was able to understand most of the things I had always hoped to understand in a surprisingly short amount of time.

What I learned was basically exactly what Nobel Prize winner I. I. Rabi had pointed out almost a century earlier:[13]

[13] As quoted in [Derman, 2004]

"If you decide you don't have to get A's, you can learn an enormous amount in college."

You might be surprised that it took me so long to discover Feynman's famous lecture notes. In hindsight, my main problem was my belief that the only way to learn a complicated topic like physics was to push through a university program. I wasted years by attempting to follow the university curriculum religiously.

If I could start over, I would do everything differently. I would focus on educating myself from the start.

But no one had told me not only was it possible to teach physics to yourself, but it was also more exciting, effective, and fulfilling. I spent years attending boring lectures and grinding through the busywork of assignments. And it was always a fight because I had to learn so many things I had no interested in, just to pass the exam. Moreover, I had always waited until someone "allowed me" to study a particular topic.

At least in my experience, this is how things work for most students. There is rarely any time for extra reading in a standard physics program. As a result, most students never read a book that isn't mandatory reading. They focus solely on getting good grades since they trust the university curriculum and want to increase their career prospects. I witnessed a countless number of my fellow students go from passionately curious to demotivated. After a few years, most of them simply accepted that they would never understand things properly.

But it's not only that most people don't know it's possible to properly learn physics themselves. It's also really hard to carve out an alternative path completely on your own, without any guidance. I probably would have buried my dream of understanding physics if I hadn't found Feynman's lectures by chance in the library. Although things got simpler over time, self-learning physics was never easy because I didn't have a map or handbook.

Therefore, my goal is not only to motivate others to teach physics to themselves, but also provide them with the tools and knowledge to do so. That is why I wrote this book.

Disclaimer

I'm the author of several textbooks and, where appropriate, I will mention them in the text below.

And to be honest, I was somewhat unsure whether or not I should do this.

On the one hand, I'm genuinely convinced that they are helpful — otherwise, I would not have spent years of my life writing them. And just imagine if the author of a classical mechanics book recommended several other classical mechanics books but never mentioned his own. That would be quite strange.

However, I certainly don't want readers to think that I only wrote this book to promote my other books.

While I will highlight books that I wrote myself, I will also recommend multiple alternatives. My textbooks are certainly not perfect and there is no one-size-fits-all approach when it comes to understanding.

I also have zero affiliation with any of the other books I recommend below. The authors are not friends of mine and I don't get any money (or any other form of compensation) for recommending their books. I only recommend books that I genuinely think are helpful for students at different stages of their journey.

Contents

1 **The Plan** 29

Part I Why Should You Teach Yourself Physics?

2 **Why Should You Teach Yourself *Physics*?** 35

3 **Why Should You Teach *Yourself* Physics?** 39
3.1 The Joy of Backpacking 41
3.2 Hill Climbers and Valley Crossers 49

4 **Why Should *You* Teach Yourself Physics?** 53

Part II Overview

5 **Bird's Eye View of Physics** 61

6 **The Structure of Physics** 65
6.1 The Cube of Physics 68
6.2 Why Physics Works 71
6.3 Formulations and Interpretations 76

Part III Subject Guides

7 Classical Mechanics **87**
7.1 Why Learn Classical Mechanics? 88
7.2 Bird's Eye View of Classical Mechanics 89
7.3 How To Understand The Fundamentals 92
7.4 Map of Classical Mechanics 94
7.5 How To Level Up 94

8 Electrodynamics **97**
8.1 Why Learn Electrodynamics? 98
8.2 Bird's Eye View of Electrodynamics 100
8.3 How To Understand The Fundamentals 104
8.4 Map of Electrodynamics 106
8.5 How To Level Up 107

9 Special Relativity **111**
9.1 Why Learn Special Relativity? 112
9.2 Bird's Eye View of Special Relativity 113
9.3 How To Understand The Fundamentals 117
9.4 Map of Special Relativity 117
9.5 How To Level Up 118

10 Quantum Mechanics **121**
10.1 Why Learn Quantum Mechanics? 122
10.2 Bird's Eye View of Quantum Mechanics 123
10.3 How To Understand The Fundamentals 127
10.4 Map of Quantum Mechanics 129
10.5 How To Level Up 129

11 Quantum Field Theory **133**
11.1 Why Learn Quantum Field Theory? 134
11.2 Bird's Eye View of Quantum Field Theory . . . 135
11.3 How To Understand The Fundamentals 138
11.4 Map of Quantum Field Theory 141
11.5 How To Level Up 142

12 General Relativity 147
12.1 Why Learn General Relativity? 148
12.2 Bird's Eye View of General Relativity 149
12.3 How To Understand The Fundamentals 151
12.4 Map of General Relativity 153
12.5 How To Level Up 153

13 Statistical Mechanics and Thermodynamics 155

Part IV Further and Meta Subjects

14 Learn How to Learn 163

15 Learn How to Think 165
15.1 Biases and Fallacies 167

16 Mathematics 173

17 History and the Big Picture 179

18 Philosophy 183

19 Sociology 189

Part V Advice and Principless

20 One Thing You Must Understand 201

21 The Master Key 207
21.1 Teach 210
21.2 Jump 217

22 Practice 219

23 Time 221

24 Beyond Dogma 223
24.1 Physics Beyond Dogma 226

24.2 Leveling Up 228

25 Closing Words **231**

One Last Thing

Part VI Appendix

A FAQ **237**

Bibliography **243**

1

The Plan

This book is a paradoxical project. On the one hand, I want to encourage you to think independently, to stop following predefined paths, and to discover things on your own terms. On the other hand, I want to offer guidance and give specific recommendations, since even the most independent self-learner sometimes needs some guidance. The solution I came up with is that I will give many recommendations and share lots of opinions. However, I will remind you several times that everything written here is just my personal perspective, and you should always consult other sources since one perspective is never sufficient.

To give you a place to start, here are a few alternative roadmaps and lists of book recommendations curated by different people:

- *How to Learn Math and Physics,* by John Baez-[1]
- *So You Want to Learn Physics...,* by Susan Fowler.[2]
- *How to become a GOOD Theoretical Physicist*, by Gerard 't Hooft.[3]
- *The Chicago Undergraduate Physics Bibliography*, curated by Abhishek Roy.[4]
- *The Physics Book Recommendations* at Stackexchange.[5]
- *So You Want to Become a Physicist?*, by Michio Kaku.[6]
- *Math and Physics Book Recommendations* by Spencer Stirling.[7]

You can use these resources, plus my recommendations, to build a learning roadmap that perfectly fits your needs and interests. Arguably the most important skill you will learn while studying physics is the ability to judge for yourself what will help you move forward.

But don't become a prisoner of your roadmap. To learn physics effectively, you need to always adjust your roadmap as you move along. It's simply impossible to create a perfect roadmap in the beginning and then rigorously work your way through it. And, if you're waiting until you have all the information, you'll find yourself at a standstill.

[1] http://math.ucr.edu/home/baez/books.html

[2] https://www.susanjfowler.com/blog/2016/8/13/so-you-want-to-learn-physics

[3] http://www.goodtheorist.science/

[4] https://www.ocf.berkeley.edu/~abhishek/chicphys.htm

[5] https://physics.stackexchange.com/questions/12175/book-recommendations

[6] https://mkaku.org/home/articles/so-you-want-to-become-a-physicist/

[7] http://www.spencerstirling.com/mathgeek/mathnotes.html

Creating a roadmap is essential because it allows you think about the journey as a whole and how things fit into context. But you should accept that your roadmap will always be terribly flawed and will need to be constantly updated. "Plans are worthless, but planning is everything", as Dwight D. Eisenhower famously remarked.

With that said, let's briefly discuss what we'll talk about in the next chapters.

First of all, we discuss why people care about physics at all. To learn anything effectively you need to have a proper reason why. Otherwise, learning quickly turns into a joyless exercise of jumping through hoops.

After that we will spend some time thinking about the journey as a whole and develop some understanding of the "big picture". Where are we going? And what are we trying to achieve?

Then we will be ready to get specific. Which subjects should you learn to get a proper understanding of physics? And how can you learn them most effectively?

Let's dive in.

Part I
Why Should You Teach Yourself Physics?

"We shall not cease from exploration, and the end of all our exploring will be to arrive where we started and know the place for the first time."

T. S. Eliot

2

Why Should You Teach Yourself *Physics*?

Physics is one of the most exciting intellectual adventures because it allows us to tackle really big questions:

- ▷ What are the basic building blocks of nature?
- ▷ How did the universe begin, and how will it evolve?
- ▷ What is space and time?

The good news for anyone interested in physics is that there remains so much yet to be discovered and figured out. No matter how hard anyone tries to convince you, know that physics is far from being a finished book. There are dozens of known, fundamental, yet unsolved problems and quite likely even more that we don't even know of. So, if you want to get involved, there is a big chance that you will be able to help humanity advance in our communal pursuit to understand nature. And, as Paul M. Dirac once remarked:[1]

> "Living is worthwhile if one can contribute in some small way to this endless chain of progress."

[1] Graham Farmelo. *The strangest man : the hidden life of Paul Dirac, quantum genius.* Faber and Faber, London, 2009. ISBN 9780571222865

It's undeniable that new insights in physics have led to huge technological revolutions and is doing a lot in powering the world economy. For example, physics gave us nuclear power, the semiconductor, and the World Wide Web. Admittedly, the step from a discovery in physics to applications in technology sometimes take a few decades. Currently, for instance, no one has an idea what quantum field theory will allow us to do. But historically, even arcane ideas like special relativity proved to be invaluable for technological breakthroughs.[2] So it seems plausible that this will continue to be the case.

[2] GPS, in this case.

Frédérick Bordry, the CERN Director for Accelerators and Technology, recently summarized this point of view perfectly:[3]

[3] https://www.bbc.com/news/science-environment-46862486

> "When I am asked about the benefits of the Higgs Boson, I say 'bosonics'. And when they ask me what is bosonics, I say 'I don't know'. But if you imagine the discovery of the electron by J. J. Thomson in 1897, he didn't know what electronics was. But you can't imagine a world now without electronics."

Similarly, there is the (likely apocryphal) story of when Faraday demonstrated an electromagnetic induction experiment at the Royal Institution in London. He was asked, what use is it? And he replied, "What use is a baby? It grows up."[4]

[4] https://philosophy.stackexchange.com/a/61134

To really bring this point home, let's go crazy for a minute.

Imagine that some extraterrestrial civilization discovers our beautiful blue planet and — for whatever reason — decides to attack us. Who would win in such a war?

A civilization that understands more about the physics laws of nature, and how to use it for its own good certainly has an unfair advantage. In other words, the civilization that knows more physics would probably win.

Of course, to quote Niels Bohr, "It's hard to make predictions, especially about the future". So I won't try to guess what "superpowers" a more advanced civilization could have. But just imagine what a fight would look like between a civilization that knows nothing about electrodynamics and a civilization that has fully mastered it. It's reasonable to think that a fight between us and a civilization that has fully harnessed quantum gravity would look similarly or even more unfair.

I'm not proposing this thought experiment because I think such a war is going to happen anytime soon. Instead, I like to present it because it allows us to think in clear terms about what progress on the grand scale of things really means. "Science is the only news", as the writer Stewart Brand noted.

Compared to advances in science, developments that happen in politics and celebrity gossip are merely drops in the ocean. The long-term progress of our civilization and the advance of human welfare depends crucially on how much progress is made in science. Science, and physics in particular, are responsible for building up a knowledge base for all of humanity. And if this isn't reason enough to study physics, what is?

Of course, it is also worth mentioning that tools developed

in the context of physics are often invaluable in other scientific disciplines too. So even if you don't want to spend your time thinking about nature, learning physics is still a worthwhile endeavor because it teaches you a large set of tools that you can use in a broad range of contexts.

And last but not least, physics is incredibly fun and fulfilling. Once you fall in love with physics, you'll never be bored again. There are so many exciting, deep, and intellectually challenging ideas to explore that it's impossible to ever finish learning physics. But most likely, you already know why you want to study physics. Otherwise, I assume you wouldn't be reading this book.

So, next let's talk about something that is probably a lot less obvious.

3

Why Should You Teach *Yourself* Physics?

Once you've decided to learn physics, you need to think about how you want to go about it. Roughly, your choices are that you can either enroll in some university program or you can try to teach it to yourself.

While participating in a university program is pretty self-explanatory, you might wonder what it really means to teach *yourself* physics?

In short, it all boils down to Richard Feynman's recommendation:

> "Study hard what interests you the most in the most undisciplined, irreverent and original manner possible."

So, in other words, you ignore all predefined structures and try to carve out your own path. Of course, this can mean that you follow a university curriculum (online or offline) for a while. Or maybe you just pick a few lectures that fit your needs. Or you ignore universities completely and focus on books and other resources instead. We'll talk about such details later.

Here, we'll only discuss why — whether you're fresh out of high school, enrolled at university, or already retired — you must teach physics to yourself to develop a truly deep understanding of it.[1]

[1] It also doesn't matter at what point of your journey you're currently at. It's never too late to change your approach.

3.1 The Joy of Backpacking

I like to think of physics as a huge continent. There are both beautiful places and quite ugly ones — although not necessarily everyone agrees which is which. There are mountains you can climb (e.g. "Mount Symmetry") which allow you to see incredibly far. And there are certain landmarks (e.g. Schrödinger's cat) everyone knows about.

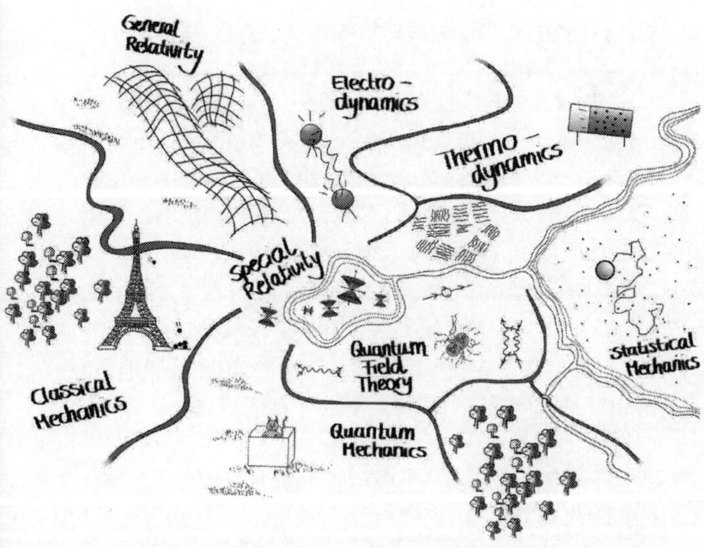

To truly understand what is going on throughout the continent you need to learn its language, which in this case is called mathematics. Furthermore, one aspect that makes this continent so exciting is that there is still lots of terrain no one has ever explored. As Isaac Newton once remarked:

> "The greater the continent of knowledge; the greater the coastline of the unknown."

The analogy is especially useful if we want to discuss different ways to explore the continent of physics. You can, for example, book a fully-guided bus tour (a.k.a. enroll in a standard university program) or you can start exploring on your own terms.[2]

[2] Both alternatives (and everything in-between) are, of course, viable options. For some people, a fully-guided bus tour might be the right choice. But if your goal is to feel truly at home in the continent of physics, it's essential that you explore it independently.

Just like in the real world, backpacking through the continent of physics will not only provide a far more satisfying experience, but will also give you the confidence that you need to start exploring uncharted territory.

Let me explain in bit more detail why.

First of all, a normal university program follows, just like a guided bus tour, a completely fixed timeline. Everyone spends exactly the same amount of time learning about, say classical mechanics and listens to exactly the same explanations. This is problematic since every student brings a unique background to the table and therefore may need a bit longer or shorter to truly understand a given concept. With a fixed curriculum students are regularly forced to move on even though many questions remain unanswered.

The fixed timeline is not only problematic because it forces different students to spend exactly the same amount of time on a given topic, but also because usually the same exact amount of time is allocated for different topics.

Just think about it. What are the chances that it's the perfect approach to spend exactly the same amount of time on completely different topics like classical mechanics, quantum mechanics or electrodynamics? As crazy as it may sound, this is the standard approach at almost all universities. In contrast, as a self-learner, you're free to spend, for instance, two months learning classical mechanics, four months learning electrodynamics, and eighteen months learning quantum mechanics. And you're also free to spend, let's say, one month learning classical mechanics, one month learning electrodynamics, one month learning quantum mechanics, and then again two months learning classical mechanics, two months learning electrodynamics and so on.

In a normal university program students are expected to learn topics like classical mechanics exactly once for a fixed amount of time and then be done with it.

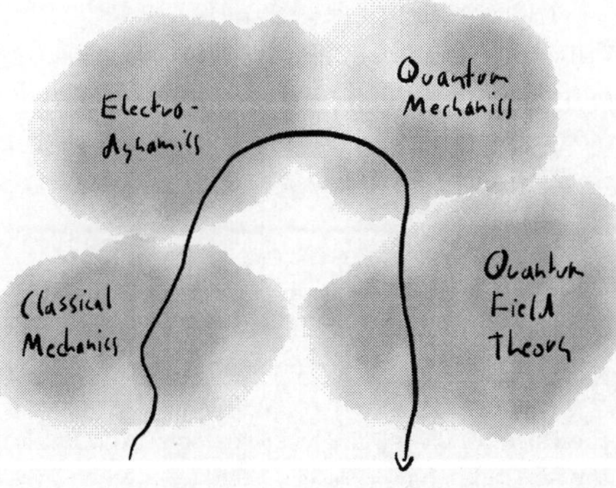

But that's not how anyone really understands physics. Everyone with a truly deep understanding of physics has revisited all fundamental topics dozens of times at different stages in their learning journey. This is necessary because as your whole understanding of physics expands, you will be able to appreciate completely new aspects of a theory like classical mechanics.

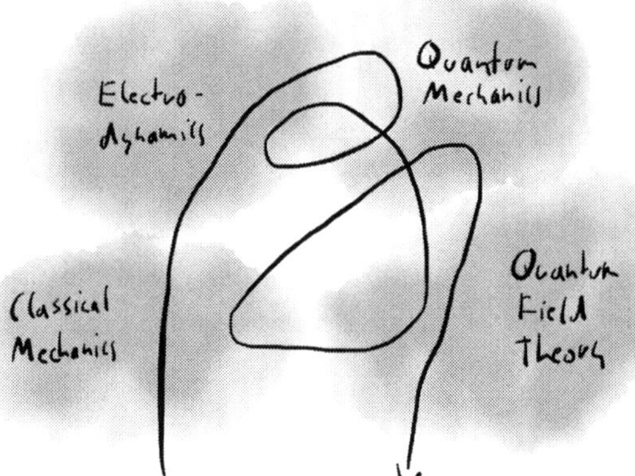

Secondly because of their backgrounds, students typically look at a given topic from a very personal perspective and therefore, find quite different explanations illuminating. This is nicely illustrated by the following story Andrew Low told in his lectures on finance theory in the fall of 2008:

> "You're going to have to focus on two concepts that are probably alien to you. The notion of a stock and a flow. Now when I say stock, I don't mean common stock or equities, I have a different term in mind. By stock in this context, I mean the stock of assets. The level of assets. And by flow I mean the rate of change of assets. You know when I was in grad school, we started discussing this concept on the first day of macroeconomics, and then one of the students in the back of the room said, excuse me, Professor, but isn't that just the distinction between a variable and its first derivative? And the professor was a little bit taken aback and said, well, yes, that's right. But let me give you another way of thinking about it that is somewhat more intuitive. And that is, think about a bathtub, and think about the faucet turned on and the water flowing into it. The stock is the level of the water. The flow is how fast the water is coming into the tub. And so after that explanation the student still seemed confused, and so the professor said, you know what, some people find bathtubs intuitive, other people find derivatives intuitive. So

to each his own."

This little anecdote nicely exemplifies how there can be huge differences in what kind of explanation different people find illuminating.

Another great example is the seven ways to understand derivatives listed in William P. Thurston's amazing essay "On proof and progress in mathematics". Some student's find it illuminating to think of a derivative in terms of "the ratio of the infinitesimal change in the value of a function to the infinitesimal change in a function", while others might find it illuminating to hear that "the derivative of a real-valued function f in a domain D is the Lagrangian section of the cotangent bundle $T^\star(D)$ that gives the connection form for the unique flat connection on the trivial R-bundle $D \times R$ for which the graph of f is parallel."[3]

[3] William P. Thurston. On proof and progress in mathematics, 1994

To understand any topic properly, it's never sufficient to just listen to some lecture. You need to critically engage with the content. You need to raise questions and then search for answers no matter how long it may take. But there is no room for this process within the rigid structures of a pre-defined curriculum. This is analogous to how during a guided bus tour, there is usually only time for one or two quick snapshots and, if the guide is in a good mood, one or two superficial questions before the group has to move on to the next sight.

However, self-learners can spend as much time as they need on any given topic and can build a curriculum that perfectly matches their interests.

The odds that the one local guide traveling with you explains everything in understandable terms and does it at exactly the right pace is close to zero. Moreover, your local guide may or may not be skillfull at explaining things. But most certainly, he or she will only be able to present one perspective. In a guided tour, there is usually no room for variety. The company provides one guide and that's that. They've done their duty.

However, if you actively teach physics to yourself, you can make sure that you learn from many skillful teachers. This is possible because many great teachers have written excellent books. Therefore, you can learn from the best physics teachers that have ever lived no matter where in world you currently are.

To make matters even worse, the curricula at universities are usually not designed with the primary goal to help their students understand physics inside out and don't prepare them well to do actual science.[4] In our analogy, we could say that a university program prepares you for a job in the tourism industry but not for a life as an explorer.

[4] Or, even more extreme, as Sanjoy Mahajan puts it: "Traditionally taught science and mathematics teach little except obedience." [Mahajan and Hake, 2005] If you're puzzled by these kinds of remarks, you might enjoy reading Chapter 13 in Ref. [Simler and Hanson, 2018] and Ref. [Schmidt, 2000].

Undergraduates typically learn an enormous number of facts, several calculation methods, and basic familiarity with some experimental equipment. The learning primarily takes place through lectures in classrooms, artificial puzzle-like problem sets, and lab courses where student's are expected to push a few buttons to get the known-correct answers. However, the primary goal science is to discover previously-unknown truths and almost nothing students learn prepares them for this task.

Most university curricula are simply the product of many historical developments and would certainly look completely different if they were redesigned from scratch. Just as busses can only drive along certain paved roads, university programs all around the world follow almost exactly the same curricula.

So, it's your responsible to ignore the noise around you and explore things at your own pace. You need to stop whenever you think a pause is necessary. You always need to go as deep as you think is necessary.

This is not easy. But it's the only way to develop a truly deep understanding of physics. In the words of Isaac Asimov:[5]

> "And self-education is, I firmly believe, the only kind of education there is."

[5] Isaac Asimov. *Science past, science future.* Doubleday, Garden City, N.Y, 1975. ISBN 9780385099233

There are, in fact, a few weeks in a typical university program when students become self-learners: when they prepare for their final oral exams. During these weeks, students try to guess which questions their professors will most likely ask them and then try to find great answers in books or online. With what we've been discussing so far, you will probably not be surprised when I tell you that it's really during this time where most of the understanding happens. Almost all student's I've talked to in the past few years have reported that they learned more during these weeks of independent learning than in all previous years of spoon-fed learning combined. Most importantly, students usually rediscover at that time how much fun physics can be. They suddenly remember why they started studying physics in the first place and feel as if they had been released from a straight-jacket.

So, my modest suggestion would be that you only focus on this really effective type of learning and skip everything else.

Always remember that you're the only person responsible for your education. It's simply not a valid excuse that you don't know or understand *X* because your teacher didn't explain it properly. There are dozens of books on any given topic and chances are high that one of them will answer your question. It's your task to browse through twelve bad books until you finally find one that makes everything fall into place.

So in short, you need to carve out your own path and teach physics to yourself.

3.2 Hill Climbers and Valley Crossers

If you need further motivation, let me assure you that right now is the perfect time to be a self-learner in physics. The era of the autodidact hast just begun as a result of the limitless possibilities of the internet. In previous centuries, knowledge and books were rare. It was almost impossible to be a self-learner and universities are products of these times. But nowadays, everything is on the internet. If you want to learn something, you can watch Khan Academy videos or MIT lectures online. You can even download the coursework and get feedback from instructors. There are more amazing books available online than you could ever read and thousands of websites full of explanations and illuminating discussions.

Additionally, now is the perfect time to teach yourself physics because there is a need for educated, independent thinkers more desperately now than ever before. There has been little to no progress on fundamental physics problems during the past decades.[6] To overcome this current phase of stagnation, we need more well-educated people who are brave enough to tackle big questions, discover new connections and offer fresh perspectives.

[6] For an attempt to quantify the current lack of progress, see `https://www.theatlantic.com/science/archive/2018/11/diminishing-returns-science/575665/`. Interesting thoughts on the current state of physics can be found in Ref. [Hossenfelder, 2018] and Ref. [Giudice, 2019].

Here's a nice way to understand all of this in terms of our physics-as-a-continent metaphor which was articulated by Lee Smolin (who attributes it to Eric Weinstein):[7]

[7] Lee Smolin described the metaphor at `https://backreaction.blogspot.com/2006/08/lees-comments.html`. Similar metaphors are described by Freeman Dyson in his essay titled "Frogs and Birds" and by Michael Arrington in his essay "Are You A Pirate?" to name a few.

> "Let us take a different twist on the landscape of theories and consider the landscape of possible ideas about post standard model or quantum gravity physics that have been proposed. Height is proportional to the number of things the theory gets right. Since we don't have a convincing case for the right theory yet, that is a high peak somewhere off in the

> distance. The existing approaches are hills of various heights that may or may not be connected, across some ridges and high valleys to the real peak. We assume the landscape is covered by fog so we can't see where the real peak is, we can only feel around and detect slopes and local maxima.
>
> Now to a rough approximation, there are two kinds of scientists — hill climbers and valley crossers. Hill climbers are great technically and will always advance an approach incrementally. They are what you want once an approach has been defined, i.e. a hill has been discovered, and they will always go uphill and find the nearest local maximum. Valley crossers are perhaps not so good at those skills, but they have great intuition, a lot of serendipity, the ability to find hidden assumptions and look at familiar topics new ways, and so are able to wander around in the valleys, or cross exposed ridges, to find new hills and mountains."

Following a university program is a great way to become a hill climber.[8] And, of course, hill climbers were immensely important during the past decades when lots of details of established theories were carefully worked out.[9]

But to make the next huge jumps forward, we need a much larger number of valley crossers. A valley crosser ignores all pre-defined categories and explores routes no one else has ever explored. To be able to do this, you need a deep understanding of large parts of the landscape. You need to know unimpressive landmarks no one else seems to care about. What's more, you need a lot of confidence in your ability to explore things on your own.

As a self-learner, you are free to develop a much broader perspective and leave the well-worn paths. This approach is essential in order to develop unique understandings and generate new insights. Regular students all read the same books and follow exactly the same curriculum. But "if you only read the books that everyone else is reading, you can

[8] For a detailed discussion of why this is the case, see R[Schmidt, 2000].

[9] To use the terminology introduced by Thomas Kuhn in his famous essay "The Structure of Scientific Revolutions" [Kuhn, 2012], hill climbers are essential during phases of consolidation, in which existing theories and models are understood at much deeper levels. In contrast, valley crossers are essential for paradigm shifts.

only think what everyone else is thinking."[10] Thus, it is essential to get into the habit of actively seeking out sources that no one else is reading. The earlier you take the leap, the better.

[10] Haruki Murakami. *Norwegian wood*. Vintage, London, 2003. ISBN 9780099448822

Nothing will prepare you better for being a valley crosser than your time as a backpacker.

So in summary:

university program	self-learning
fixed timeline	flexible timeline
same explanations for everyone	explanations that perfectly match your needs
each topic discussed once	freedom to revisit each topic multiple times
one lecturer per topic who may or may not be a great teacher	learn from the best teachers that have ever lived
everyone learns the same concepts and perspectives	develop unique perspectives

Of course, everything said here only applies if your primary goal is to understand physics. If instead, your goal is to get a physics degree as a stepping stone for a career, you absolutely need to follow the curriculum of your local university religiously and ignore most of what I just said.[11]

But more on this in the next chapter.

[11] In the future there might be "college equivalence degree" that you are awarded if you can prove that you have the equivalent knowledge of a college graduate. But since these do not exist yet, universities are still the best way to earn credentials as a career stepping stone.

4

Why Should *You* Teach Yourself Physics?

One final thing we need to talk about before we really dive into the meat of the matter: Why *you* in particular should teach physics to yourself. Well, maybe you shouldn't.

This book is certainly not a good fit for everyone. This book is probably not the right choice if your primary goal is to make a career in physics. As most professors will happily confirm, a burning desire to understand things deeply can often be a hindrance for your career.

"Shut up and calculate" is often a much better career advice than what I can offer in this book. To make a career in physics in the current academic system you need, first of all, good grades. Otherwise, you will not be able, as one example, to find a PhD position at a top university. And to get good grades, you need to learn how to solve exercises quickly and without errors. The more you practice, the better you get. And each hour spent contemplating the meaning and origin of various concepts, is an hour you could've spent practicing how to solve integrals instead.

Of course, being able to solve exercises quickly and without errors by no means correlates with your ability to conduct good research. In the real world, researchers usually have all the time in the world to solve the problem they are working on. And they can always ask colleagues for help or to do cross checks, and use tools like Mathematica or textbooks to look up formulas.

However, this is simply how the system currently works. There are lots of students in university classes and professors need a scalable way to assess students. That's why good grades are essential for a successful career in physics.

Later in your career, it's not just "shut up and calculate" but also "publish or perish". To find a research position — maybe even a permanent one — you need be able to write and publish a lot of papers that get a significant number of citations in the first few years after their publication. If you "waste" your time contemplating deep questions or trying to come up with novel connections, you will probably never make it very far in the current system. Progress on deep questions and the discovery of novel connections are simply not plannable and during the first few years most other researchers will probably not understand what you're talking about. Therefore, this is not a good way to boost your h-index.[1]

In summary, if you spend some time in the physics community you will find, to quote Lee Smolin, the following:[2]

[1] For a much more thoughtful take on this issue see Smolin's excellent "Why no new Einstein?" Ref. [Smolin, 2005].

[2] https://www.psychologytoday.com/intl/blog/finding-the-next-einstein/201309/lee-smolin-encourages-graduate-student-stay-in-science

> [T]here is, to put it simply, an ongoing fight between those of us who do science to satisfy our curiosity about nature and increase our knowledge and those who do it for careerist or egotistical reasons.[3]

If you belong to the latter category, you will find that books like the following are much better suited for your needs:

▷ A PhD is not enough by Peter Feibelmann [4] and

▷ The Professor Is In by Karen Kelsky [5]

These books are perfect companions to survive in the jungle that we call the modern academic system.

But if your goal is to learn physics to satisfy your curiosity about nature, I'm confident that you you've picked exactly the right book and self-learning is the way to go. Since the modern academic system is so competitive and mainly focuses on things like grades and publications, it's inevitable that you take your education into your own hands.

Even if you're solely motivated by your curiosity about nature, you might be concerned whether or not you have what it takes to self-learn physics. Maybe you think you're simply not smart (or talented or whatever excuse you want to put in her) to truly grasp the fundamentals of physics?

I've already explained before why I'm convinced that such statements are harmful nonsense. But since such beliefs are so widespread, I want to elaborate on this point.

[3] This observation is, of course, not new. Already Albert Einstein remarked that "[o]f all the communities available to us, there is not one I would want to devote myself to except for the society of the true seekers, which has very few living members at any one time."

[4] Peter Feibelman. *A PhD is Not Enough! : A Guide To Survival in Science*. Basic Books, New York, 2011. ISBN 9780465022229

[5] Karen Kelsky. *The professor is in : the essential guide to turning your Ph.D. into a job.* Three Rivers Press, New York, 2015. ISBN 978-0553419429

First of all, let me assure you that if you don't understand something, you simply haven't found an explanation that speaks a language you understand yet. It's that simple.[6] Different people have different foreknowledge and have had different experiences in the past. For instance, maybe one person had a teacher in high-school who talked for months about nothing but hydrodynamics. As a result, all explanations that map new concepts into the language of hydrodynamics will be immensely helpful for that person but completely useless for others who have never studied hydrodynamics. Let me assure you that it's completely wrong to think that you need to be a certain type of person to succeed in physics. No one can predict what skill, insights and experiences will lead us to the next breakthrough discovery in physics. It's nonsense to think that you need to spend your best years doing complicated calculations to prepare for whatever may come in the future. It's nonsense to think that you must be capable of doing the most complicated calculations before you can add something. It's nonsense to think that you need to master every mathematical aspect before you can contribute anything significant to the field. Novel, deep insights can originate from everywhere:

[6] We will discuss this in more detail in Chapter 20.

- From a rigorous proof,
- from a long and complicated calculation,
- from a simple thought experiment.

It's not too hard to find examples in each category:

- t'Hooft's work on renormalizability was a rigorous proof that helped us immensely in understanding quantum field theory.
- Onsager's solution of the Ising model was a "tour de force" calculation that is still essential for our under-

standing of phase transitions.

▷ Einstein's "elevator thought experiment" was crucial for his discovery of general relativity.

Even though it is well known that deep insights can originate from everywhere, it's not too surprising why many people still believe they're too stupid to learn physics and contribute something meaningful. Within a traditional university program, usually only one particular type of skill is valued. As a result, many students feel stupid and become convinced that they don't have what it takes. This is nicely summarized by the following cartoon based on a quote often attributed to Albert Einstein:[7]

[7] "Everyone is a genius. But if you judge a fish by its ability to climb a tree, it will live its whole life believing that it is stupid." You can find a much more beautiful version of the cartoon online. Unfortunately, the original artist is unknown and thus, to avoid copyright problems, I included this adapted version here.

The crux, is that a task as complex and creative as "discovering previously-unknown truths" can certainly not be boiled down to a formalized set of tasks students need to excel in. In the cartoon, climbing the tree is certainly a valid approach to learn something about the tree. But if we really want to understand the tree's structure and, say, how it grows, quite different skills are needed.

With this much broader goal in mind, it's clear how all of the animals in the cartoon could contribute something meaningful. The fish could possibly help figure out where the water comes from that keeps the tree alive. The elephant could use his strength to open up a "window" to look inside. Maybe the bird, acting as the "valley crosser" in this picture, discovers that there are lots of other trees and different kinds of structures we don't even have a name for yet.

To summarize, no has the right to tell you that you don't have what it takes because no one knows what skills and experiences will prove to be invaluable in the future. Many will try to convince you otherwise because they think it's fun to be part of some elite club. In the eyes of the general public, physicist are people with enormous mental capabilities and most physicists certainly have no interest in destroying that myth.

But physics is something that can and should be enjoyed by anyone motivated to do so. There is no gate, only artificial barriers some people try to construct. You should definitely come and see for yourself.

Part II
Overview

"I do not know what I may appear to the world, but to myself I seem to have been only like a boy playing on the seashore, and diverting myself in now and then finding a smoother pebble or a prettier shell than ordinary, whilst the great ocean of truth lay all undiscovered before me."

Isaac Newton

5

Bird's Eye View of Physics

In a nutshell, physics can be summarized as follows. We take a chunk of reality that we want to understand and translate it into mathematics. More specifically, this means that we encode our findings in the form of equations. Using these equations, we can make predictions and test our ideas.

Formulated differently, our goal in physics is to describe, understand, and explain what is going on in nature.

There are, of course, lots of scientific disciplines which try to make sense of nature. So how does physics fit into the bigger picture? One answer to this question can be illustrated this way:[1]

[1] The following hierarchy is adapted from the one put forward by George F. R. Ellis in his paper "Physics and the Real world".

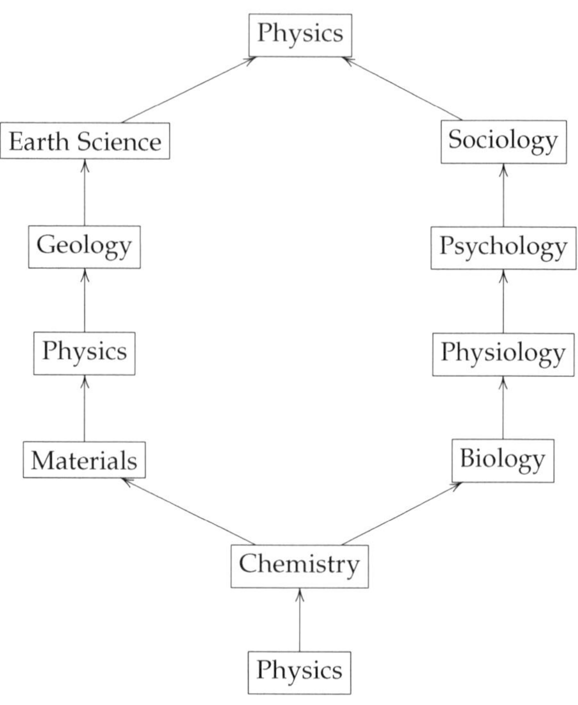

In words, this means that in physics we try to describe nature at the largest (planets, galaxies, the universe) and smallest scales (elementary particles). But these are only the grandest of goals. We can also use our physics toolbox to make sense of, say, everyday objects or electromagnetic waves. The thing is, there is not only a hierarchy of different scientific disciplines, but also a hierarchy within physics.[2] This is something we will talk about in a moment. But first, we need to talk about what it is that we do in physics or, more commonly, try to do.

[2] We will talk about the relationship between physics and other scientific disciplines in more detail in Chapter 6.2.

First of all, of course, we observe. Then, as soon as we have found some phenomena or class of phenomena we want to describe, we can start "doing physics".[3] Our task is to find suitable handles onto the world. In physics, usually we

[3] The following characterization of physics is due to Martin H. Krieger and was put forward in his book 'Doing Physics' [Krieger, 1992].

call these handles "degrees of freedom". Examples are the position and momentum of a particle, the frequency and amplitude of an oscillator, or the temperature and pressure of a gas.

Next, we try to come up with experimental setups which allow us to isolate individual degrees of freedom as much as possible. In other words, we shield a specific tiny part of the universe (our subsystem/our experiment) from irrelevant degrees of freedom, so that we can investigate the degrees of freedom we are interested in. Then, to learn something about Nature, we "poke it". In concrete terms, this means that we shake our handles and observe how they react. Martin H. Krieger describes it poetically:[4]

> "One needs to shake the handle with just the right energy, and in just the right direction, and one will hear the music of Nature in its purest forms."

[4] Martin Krieger. *Doing physics : how physicists take hold of the world.* Indiana University Press, Bloomington, 1992. ISBN 978-0253207012

We then try to describe what we observe by writing down a "model". A model consists of formulas that relate the mathematical objects representing our degrees of freedom.

After some time, there will be a large collection of models describing all kinds of phenomena related to several specific degrees of freedom. This is usually the time for a huge step forward. Someone has to come up with something that allows us to see the forest and not just the trees. In other words, the goal is now to find a "theory" that allows us to understand all these models in a common context.[5]

[5] We will talk about concrete examples of theories, models and their relationship below.

After a theory has been proposed, it must be tested in experiments. Crucially, it is not enough that the theory allows us to understand all relevant known phenomena, but it must also predict things we didn't see previously. A good theory makes predictions not only postdictions. The symbiotic relationship between experiments and theoretical ideas is one of the hallmarks of physics. In the words, of Richard Feynman:[6]

[6] Or, to quote T. D. Lee: "Without experimentalists, theorists tend to drift. Without theorists, experimentalists tend to falter."

> "It doesn't matter how beautiful your theory is, it doesn't matter how smart you are. If it doesn't agree with experiment, it's wrong."

[7] The "scientific method" and related aspects are discussed in more detail in Chapter 18.

Often, theoretical ideas are inspired by experimental observations. In turn, theoretical ideas inspire new experiments which then lead to new theoretical ideas and so on.[7]

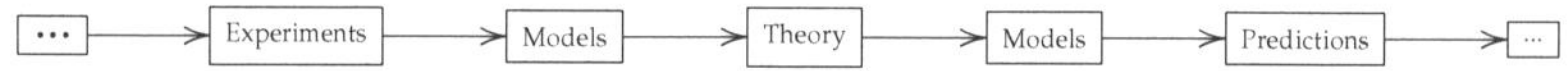

For example, physicists used collider experiments to investigate the properties of elementary particles like quarks, photons, and electrons. The behavior and interplay of these particles can be described using models like Quantum Electrodynamics and Quantum Chromodynamics. Both are models in the framework of quantum field theory. And in turn, physicists used quantum field theory to propose new models like the Electroweak Model which was confirmed experimentally through the discovery of the Higgs particle.

In the next chapter, we'll talk about the internal structure of physics and especially the relationship between theories and models in a bit more detail.

6

The Structure of Physics

In physics, there isn't one universally applicable theory. Each theory has specific strengths and weaknesses. Which theory we use always depends on the system at hand. This can be quite confusing, especially for beginning students. Many first-year students are frustrated why they have to learn an "outdated" theory like classical mechanics. Everyone knows that classical mechanics has been replaced with quantum mechanics, right?

Actually, that's not correct. Classical mechanics is still the best theory of everyday objects that we have. Quantum mechanics does not help us in any way to describe how, say, a ball rolls down a ramp. But quantum mechanics is, of course, perfectly suited for other systems, like the hydrogen atom where classical mechanics fails horribly.

In concrete terms, (oversimplifying a bit):

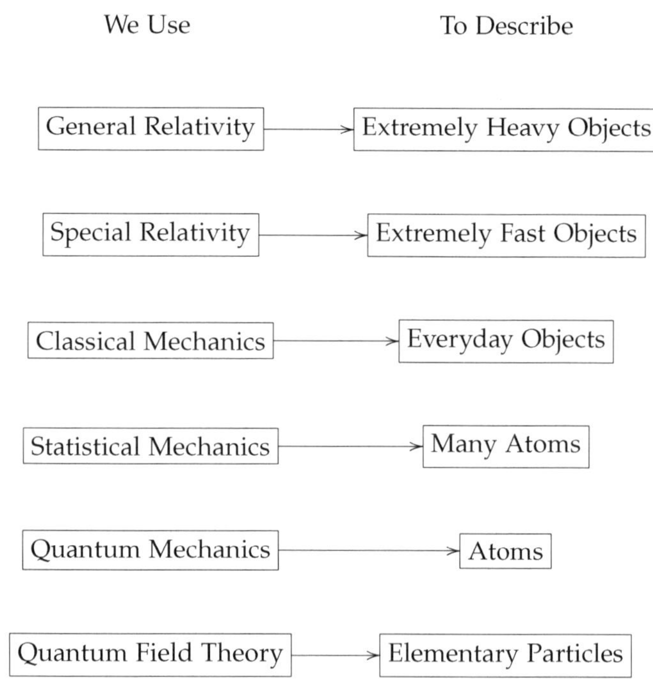

While this way of looking at the structure of physics can be helpful, there are also other ways of looking at it. For example, the following table summarizes the role of the most important physical theories and models a bit more systematically.

Theory	Model	Application	Example
Classical Field Theory	General Relativity	Cosmology	Black Holes
	Electrodynamics	Optics, Electrical Engineering	Free Electromagnetic Waves
Classical Mechanics	Special Relativity	Engineering	Length Contraction of a Moving Stick
	Newtonian Mechanics	Engineering	Ball Rolling Down a Ramp
	Statistical Mechanics	Thermodynamics	Ideal Gas
Quantum Mechanics	Non-Relativistic Quantum Mechanics	Atomic Physics	Hydrogen Atom
	Relativistic Quantum Mechanics	Atomic Physics	Thomas precession
Quantum Field Theory	QED, QCD, Standard Model	Particle Physics	Bhabha Scattering
	Ising Model, Ginzburg-Landau model	Solid State Physics	Ferromagnet, Superconductors

Whenever you feel lost while studying physics, you should come back here and see how exactly what you're learning fits into the bigger picture.

Now, let's talk about the relationship between the various theories. This can be understood nicely using an illustration known as "the cube of physics".

6.1 The Cube of Physics

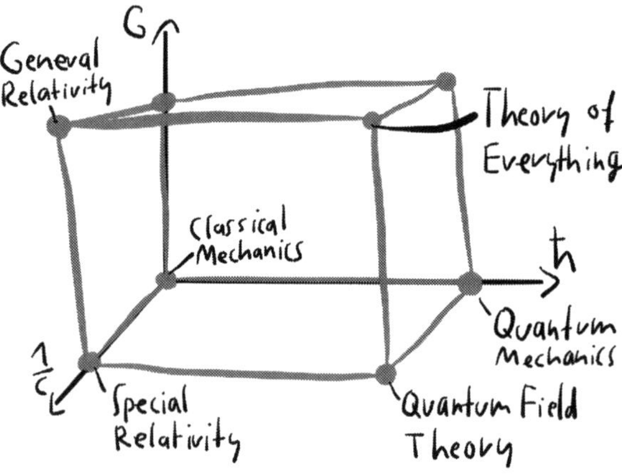

The cube of physics is a map which helps us to navigate the landscape of physical theories.[1] The main idea is to make sense of this landscape by talking about the following fundamental constants:[2]

[1] It's not entirely clear who first came up with the idea for the cube of physics, but one of the earliest appearances in literature seems to be a 1927 paper by Gamow, Ivanenko and Landau [Gamow et al., 2002]. Over the years, the cube was rediscovered many times, for example, in [Stachel, 2003].

[2] Take note that the numerical values of these constants are not really important because they depend on which units we use. If we use inches instead of meters, we get different numerical values. It is even possible to choose so-called natural units in which the numerical value of these constants is exactly 1.

▷ The speed of light $c = 2.9979 \times 10^8$ m/s, which encodes an upper speed limit for all physical processes.

▷ The gravitational constant $G = 6.6741 \times 10^{-11} \frac{\text{m}^3}{\text{kg}\cdot\text{s}^2}$, which encodes the strength of gravitational interactions.

▷ The (reduced) Planck constant $\hbar = 1.05457 \times 10^{-34} \frac{\text{m}^2\cdot\text{kg}}{\text{s}}$, which encodes the magnitude of quantum effects.

While these are, well, constants, we imagine what happens when we vary them. This is motivated by the observation that when every object in a given system moves extremely slowly compared to the speed of light $v \ll c$, we can act as if $c \to \infty$ to simplify our equations. As mentioned above, the speed of light is an upper speed limit. No object can move with a velocity faster than c. So taking $c \to \infty$ corresponds to a situation in which there is no such speed limit at all. While there is *always* this speed limit in physics, we can act as if there were none if we only consider slowly moving objects.

Just imagine there was a highway with an upper speed limit of $v_{\text{max}} = 100000000 \frac{\text{km}}{\text{h}}$ while no car can drive faster than $v \approx 300 \frac{\text{km}}{\text{h}}$. So technically there is a speed limit, but it doesn't matter, and we can act as if there was none.

Similarly, by considering the limit $G \to 0$, we end up with theories in which there is no gravity at all.[3] And by considering the limit $\hbar \to 0$, we end up with a theory in which quantum effects play no role.[4]

So the most accurate theory of physics takes the upper speed limit ($c \neq \infty$), gravitational interactions ($G \neq 0$) and quantum effects ($\hbar \neq 0$) into account. This would be a **theory of everything** and, so far, no one has succeeded in writing it down.

While this is certainly a big problem, it doesn't stop us from making astonishingly accurate predictions. Depending on the system at hand, we can act as if certain effects don't exist at all. And this is how we end up with the various theories which live at the corner points of the cube of physics:[5]

▷ Whenever it is reasonable to ignore gravitational interactions $G \to 0$ (e.g., for elementary particles), we can use **quantum field theory**.

▷ For systems in which it is reasonable to ignore quantum effects $\hbar \to 0$ (e.g., planets), we can use **general relativity**.

▷ If we can ignore quantum effects *and* gravitational inter-

[3] Again, while there are always gravitational interactions even between elementary particles, we can often ignore them because gravity is much weaker than all other kinds of interactions. So by taking the limit $G \to 0$ we consider systems which only contain objects which are so light that gravity plays no important role.

[4] While quantum effects are extremely important for elementary particles, they have no notable effects on everyday objects. Such objects consist of so many elementary particles that the quantum effects average out, and we end up with no quantum effects at all, i.e., $v_{\text{max}} \to \infty$.

[5] The remaining corner point is at ($\hbar \to 0$ and $\frac{1}{c} \to 0$) and corresponds to Non-Relativistic Quantum Gravity which, so far, is a speculative topic that we will not discuss here.

actions ($\hbar \to 0$ and $G \to 0$), we can use **special relativity**.

▷ Moreover, when it is reasonable to ignore gravitational interactions and that there is an upper speed limit ($G \to 0$ and $\frac{1}{c} \to 0$), we can use **quantum mechanics**.

▷ For systems in which we can ignore quantum effects *and* that there is an upper speed limit ($\hbar \to 0$ and $\frac{1}{c} \to 0$), we can use classical mechanics with Newton's laws to describe gravity. (The resulting model is often called Newtonian gravity.)

▷ And finally, if we can ignore quantum effects, the upper speed limit, *and* gravitational interactions ($\hbar \to 0$, $\frac{1}{c} \to 0$ and $G \to 0$), we can use non-relativistic classical mechanics without gravity.

If you still find the cube of physics confusing, here's an alternative perspective:[6]

[6] Take note that we can, of course, take gravitational effects in classical mechanics into account as long as these are not too wild (black holes, etc.) So, the distinction between classical mechanics and Newtonian mechanics is really just semantics. Here we define Newtonian mechanics as classical mechanics including gravity.

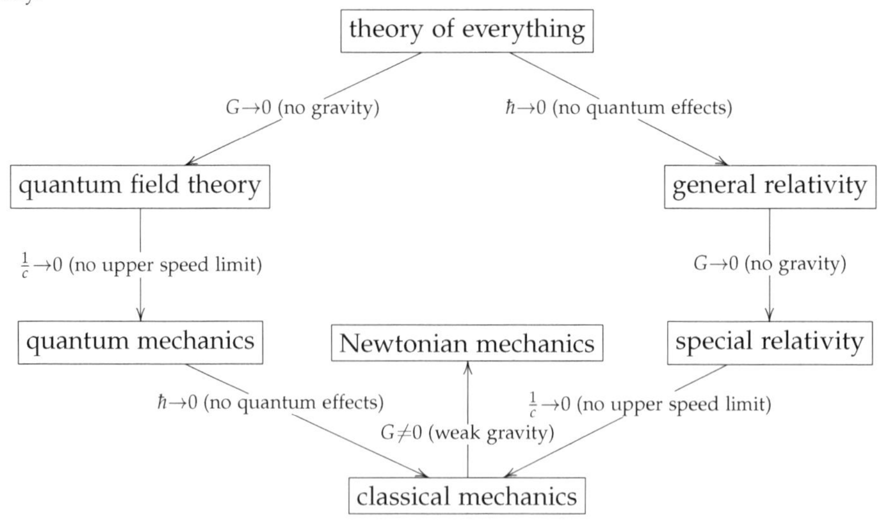

You still may wonder why people still care so much about "outdated" theories like classical mechanics. So, let's talk about this next.

6.2 Why Physics Works

> The art of being wise is the art of knowing what to overlook.
> **William James**

Why do we still talk about classical mechanics if we know about quantum mechanics? Why is classical electrodynamics still relevant if we know about quantum electrodynamics? And why do people even talk about chemistry, biology, and economics if it all boils down to physics on a fundamental level?

A pragmatic answer would be that older theories like classical mechanics or electrodynamics are necessary prerequisites for all the new stuff. Students are often told that they need to master classical mechanics before they can even hope to develop any understanding of quantum mechanics.

I personally think this is nonsense. Quantum mechanics and quantum field theory aren't intrinsically more complicated than our classical theories.[7] All theories can be extremely simple or complicated, depending on the model and application at hand. But this is not really the point I want to make here.

[7] You can learn how to draw Feynman diagrams and how to calculate the probability for different scattering processes just as quickly as you can learn how to use Newton's second law to describe classical scattering processes.

Instead, the real reason why fundamental physics hasn't taken over the world is that phenomena on different scales are usually decoupled from each other.

Let's say, for example, that we want to describe a pendulum. The basic quantities that are relevant to describe a pendulum are its period ω, its mass m, its length l and the acceleration g any object feels in the earth's gravitational field. A typical task in physics is now to predict the period ω of a specific pendulum.

To solve this problem, we first note that a period is measured in seconds, a mass in kilograms, a length in meters and an acceleration in meter per second squared. Moreover, we want a formula with the period ω on one side of the equation:

$$\omega = \_\_\_\_\_\ ? \tag{6.1}$$

Since a period is measured in seconds, the right-hand side of the formula must be measured in seconds too. Otherwise, we would be comparing apples to oranges. If we look at the basic quantities listed above, we discover that if we divide a length (measured in meters) by an acceleration (measured in meters per second squared), we get something with units second squared:

$$\text{units of } \frac{l}{g} = \frac{\text{meters}}{\frac{\text{meters}}{\text{seconds}^2}} = \text{seconds}^2 \tag{6.2}$$

Therefore, the square root of this fraction is measured in seconds and is therefore an ideal candidate for the right-hand side of Eq. 6.1:

$$\omega = C\sqrt{\frac{l}{g}}, \tag{6.3}$$

where C denotes some dimensionless constant. This equation works remarkably well.[8]

[8] As long as the pendulum only swings a little, this equation works perfectly. For larger oscillations, further correction terms are needed.

What we have discovered here is that the period of a pendulum is independent of its mass and can be predicted once its length l is known. All of this is completely independent of the pendulum's microstructural constitution. We used absolutely no details about the pendulum in the derivation above and the equation works for all kinds of pendulums, irrespective of what they are made of.

A pendulum consists of atoms which, in turn, consist of electrons and quarks. The behavior of quarks and electrons can be described using quantum field theory. But as we've just seen, such details are completely irrelevant for our description of a pendulum.

We could, of course, try to build a model of a pendulum that takes the positions, momenta, and interactions of the billions of electrons and quarks that our pendulum consists of into account. But such a project is not only deemed to fail but is also completely futile. A model of a pendulum that "employed all known physics fully" would also include the standard model, general relativity, and quantum contributions. So it would be too complex to give us any results. Therefore, any question we may have about a pendulum, can be answered perfectly using much simpler equations such as Eq. 6.3.

Also, a system that consists of so many "more fundamental" building blocks can exhibit features which simply don't exist if we only consider one or a few of the building blocks in isolation. This is known as emergence.

To illustrate this, just think about a traffic jam. A traffic jam consists of lots of cars. And cars consist of lots of individual components that consist of atoms that consist of quarks and electrons. But studying these individual components, atoms, or even quarks and electrons is not helpful if we want to understand traffic jams.

Similarly, knowing the laws of quantum field theory does

not help us describe the stock market, even though humans consist of quarks and electrons, too.

Using the language introduced in the previous chapter, we can summarize that descriptions on different scales require different handles. A description on a microscopic level must formulate its laws in terms of positions and momenta of individual elementary particles. But a description of a pendulum on macroscopic scales works perfectly well if it uses handles like its period. In between — on mesoscopic scales — even different handles may be more helpful, as demonstrated perfectly by disciplines like chemistry and biology.

The fundamental reason for this phenomenon, known as emergence, is that details become less relevant when we zoom out. Furthermore, completely new patterns can emerge, depending on how far we zoom in.

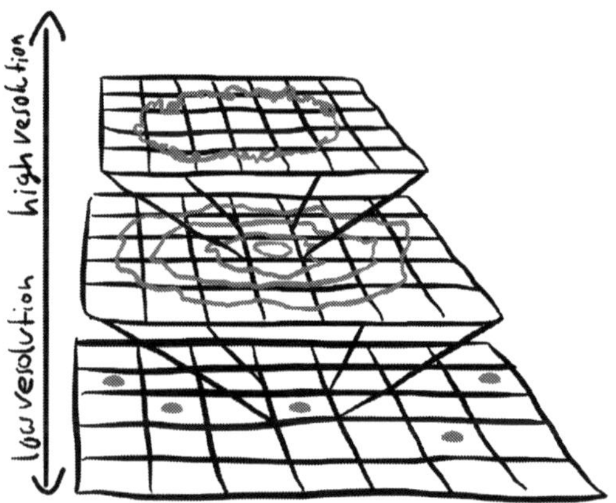

This observation is, in some sense, like a Declaration of Independence for different disciplines from each other.

In fact, it would be impossible to do physics if systems at

classical scales didn't decouple from those at quantum gravitational scales. Just imagine needing quantum gravity — a theory we know nothing about as of yet — to talk about a baseball's trajectory.

We couldn't make any predictions if microscopic details wouldn't decouple from our microscopic world since even our best theories of fundamental physics are quite likely not the end of the story. It would be astounding if our somewhat naive pictures of spacetime and elementary particles would continue to make sense at arbitrarily small scales. Every time physicists successfully "zoomed deeper in" a whole new world opened up. First quantum mechanics, then quantum field theory; and who knows what will come next.

The reason that we can use the models and theories that we currently have to make astonishingly accurate predictions is that they are independent of whatever happens at much smaller scales. We say the patterns that we observe at a given scale *emerge* from the microscopic substructure. And most importantly, the features of the handles that we use at a given scale are robust and do not depend on microscopic details. Otherwise, they would be quite useless.

Just as it would be impossible to make predictions about a baseball if its behavior depends on the exact behavior of all the atoms it consists of, it would also be impossible to describe electrons and quarks accurately if they depend on the exact behavior of whatever more fundamental stuff exists at even tinier scales.

In technical terms, we say that our theories are "effective theories" that are perfectly valid descriptions at a given scale. At some point, if we zoom in or out, our effective description becomes invalid or useless and must be replaced by a new effective description.[9]

[9] Just for the record: the mathematical tool that allows us to relate effective descriptions that are valid on different scales is known as the renormalization group.

So the key message to take away is that the various models and theories that have proven to be highly successful in the past will continue to be useful. New theories rarely replace existing ones but merely allow a description on different scales.

6.3 Formulations and Interpretations

One thing which can make physics confusing is that there are not just different theories that we need to use at different scales, but also different ways of how we can use a given theory to describe a system.

This is possible because there are different mathematical arenas we can use as the stage on which physics happens. The simplest example is a straight-forward abstraction of the physical space we live in. But there are also more abstract ones like configuration space, phase space, and Hilbert space.[10] Each of these mathematical arenas has particular advantages and, once more, it depends on the system at hand which one we use.

[10] Don't worry if these concepts are completely new to you. You will get used to them over time and there is no need to spend a lot of time thinking about them right at the beginning. For the moment, the only important thing is that you understand that these different mathematical arenas exist and how they are related to the different ways we can describe a given theory.

For example, typically quantum mechanics is formulated in Hilbert space. But it is equally possible to formulate it in phase space, configuration space or even physical space. In general, we call the description of a given theory in a particular mathematical arena a **formulation** of the theory. So in other words, there are always different formulations of each theory. This is similar to how you can describe the number 1021 using the English word "one thousand twenty-one", using the German word "Eintausend Einundzwanzig", "MXXI" in Roman numerals, or "1111111101" in the binary number system. Each of these descriptions has particular advantages depending on the context.

The following table summarizes the names of the most famous formulations of our theories:

Theory|Arena	Everyday Space	Configuration Space	Phase Space	Hilbert Space
Classical Field Theory	Standard Formulation	Lagrangian Formulation	Hamiltonian Formulation	
Classical Mechanics	Newtonian Formulation	Lagrangian Formulation	Hamiltonian Formulation	Koopman–von Neumann Formulation
Quantum Mechanics	Pilot-Wave Formulation	Path-Integral Formulation	Phase-Space Formulation	Wave-Function Formulation
Quantum Field Theory	Pilot-Wave Formulation	Path-Integral Formulation		Canonical Formulation

Disclaimer: Take note that the name "Standard Formulation" is not a term people really use. There is no standard name for the physical space formulation of classical field theory. If we simply say classical field theory or use a particular model like electrodynamics, we automatically mean the formulation in physical space. Only the formulations in more abstract spaces have a special name. Also take note that there are two empty spots which means that either these formulations have not been worked out yet or they are extremely well hidden in physics literature. Lastly take note that the pilot-wave formulation of quantum field theory has not been fully worked out yet and is not generally accepted as satisfactory.

A second crucial thing to understand is that in addition to these various formulations there are also different **interpretations** of the various theories. This is something people can get really passionate about. While no one is questioning the validity of the different formulations listed above, as they can be shown to be mathematically equivalent, people love to argue about the pros and cons of their favorite interpretations. This is especially true in the context of general relativity[11] and quantum mechanics[12].

Unfortunately, people often fail to keep the notions "theory", "model", and "interpretation" sufficiently separate. And this is something which can quickly lead to a lot of confusion.

There is one more thing I want to mention before we move on. The main reason that physics often appears so intimidating is that applications of any theory quickly become extremely complicated. The key here is to recognize that it's only the applications which are difficult, and not the theory itself. So, whenever you feel lost, try to focus on the fundamentals. This way you can always get 80% of the benefits without struggling for months trying to understand confusing mathematical machinery.

At this point it is hopefully clear that physics has an incredibly rich menu to offer and no single person can master it all. Nevertheless, it is possible to develop a deep understanding of how nature works at fundamental levels in a relatively short period of time. The key is, of course, to focus on the right things and ignore all the fluff. This is what we will talk about in the next part.

[11] In general relativity one of the main interpretational questions is the meaning of the equivalence principle. For example, it is possible to turn the standard interpretation that 'gravity = curvature of spacetime' completely upside down and argue that there is no spacetime only the gravitational field. This is known as the 'relational interpretation.

[12] There is no way to list all proposed interpretations of quantum mechanics since there are so many of them. Among the most popular ones are the 'Copenhagen formulation', the 'many worlds interpretation', the 'relational interpretation' and the 'QBism interpretation'. Also, take note that there are also debates about the interpretation of statistical mechanics, c.f. [Sklar, 1993]. One fascinating question in this context is if entropy can really be interpreted as the observer's uncertainty about the microstate [Rohilla Shalizi, 2004]. In addition, David Mermin's observation is noteworthy:[Mermin, 2019]: "Physicists have no interest in the interpretation of classical mechanics. That's part of the problem."

Part III
Subject Guides

"It is important to view knowledge as sort of a semantic tree — make sure you understand the fundamental principles, i.e. the trunk and big branches, before you get into the leaves/details or there is nothing for them to hang onto."

Elon Musk

How should you learn physics? Which subjects should you study? What is the best book for each subject?

This is what the remainder of this book is about.

First of all, anyone competent in physics should understand the fundamentals of the following eight subjects: classical mechanics, electrodynamics, special relativity, thermodynamics, quantum mechanics, quantum field theory, general relativity, and statistical mechanics.

You should focus on these topics because they are the pillars that everything else in physics rests upon. Furthermore, while studying these topics, you should make a conscious effort to focus on the fundamentals first. This is sensible strategy because while in the beginning things may look quite simple, it's easy to feel overwhelmed once you start learning about all the details and nuances. But once you've understood the fundamentals things start to look simple again.

The first stage is characterized by a naive sense of simplicity. After a while, a complexity view on the topic starts to set emerge. And only by understanding the fundamentals you can reach the final stage which is characterized by an informed sense of simplicity.[13]

[13] This model was invented by the architect Matthew Frederick [Frederick, 2007].

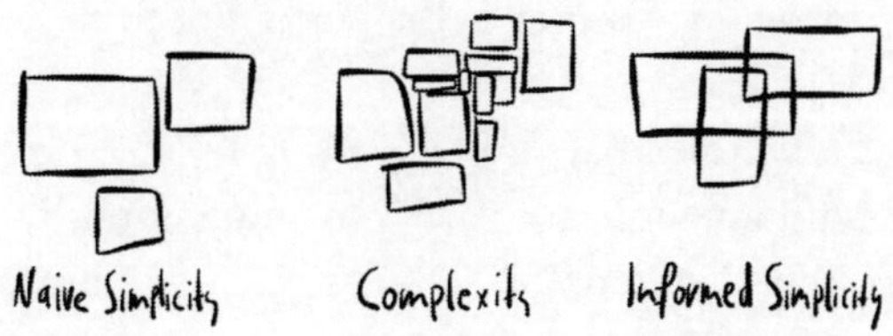

Focusing on the fundamentals will not only allow you to get a great overview of large parts of the landscape but also yield a solid foundation for everything that comes afterwards. As the classic analogy goes, if you want to "fill your cup to the brim", get your big rocks in place before your pebbles, and your pebbles in place before your sand. In my experience, the most common reason students aren't getting to where they want to be is that they focus on the wrong things; they can't differentiate between big rocks, pebbles, and sand.

In the context of physics this means that you should spend most of your time thinking about fundamental aspects like Schrödinger's equation or the electromagnetic potential and much less time on advanced applications of a particular theory.

It is especially important to ignore whatever cutting-edge research, hot topic or hype is being promoted in the newspapers and on the arXiv[14] right now. You can spend years of your life studying details of some arcane model and never get anywhere. No one knows if the proposed model has anything to do with things in the real world. What's more, within any new field, it is usually a lot less clear what is really fundamental, what the crucial assumptions are, and what is really true.

[14] The arXiv is a website (`www.arxiv.org`) physicists use to publish preprints, i.e. early versions of their papers before they get published in a journal.

Usually, your time is better spent studying the fundamentals over and over again. In other words, don't try to be on top of things; instead try to get to the bottom of things.

You might be afraid that you're wasting a lot of time doing this. The fundamentals are perfectly understood by the experts, so what's the point? There surely isn't anything new to be learned here, right?

This couldn't be any further from the truth. Let me give you just one example.

One of the most important fundamental objects in physics are spinors. We need spinors to describe, for example, electrons. But even though this is widely known, there are still lots of mysteries surrounding spinors. To quote Michael Atiyah, winner of the Fields Medal and widely regarded as one of the most influential mathematicians of the last century,[15]

[15] As quoted in Ref. [Farmelo, 2009].

> "No one fully understands spinors. Their algebra is formally understood, but their geometrical significance is mysterious. In some sense they describe the "square root" of geometry and, just as understanding the concept of $\sqrt{-1}$ took centuries, the same might be true of spinors."

So, there are definitely still many new things we can discover if we focus on the fundamentals. And you can be sure that whatever you learn about the fundamentals will definitely be relevant for our understanding of the real world. Time spent on the fundamentals is never wasted.

In which order you approach the various topics and how much time you spend on each of them is entirely up to you. But I would like to suggest either a bottom-up or a top-down approach.

The bottom-up approach is the standard route that is used by universities all around the world and also how I learned physics. It roughly follows historical developments. Here you start with the oldest theory, classical mechanics, and then move on to electrodynamics, special relativity, thermodynamics, and quantum mechanics. Afterwards, you study quantum field theory, general relativity, and statistical mechanics.

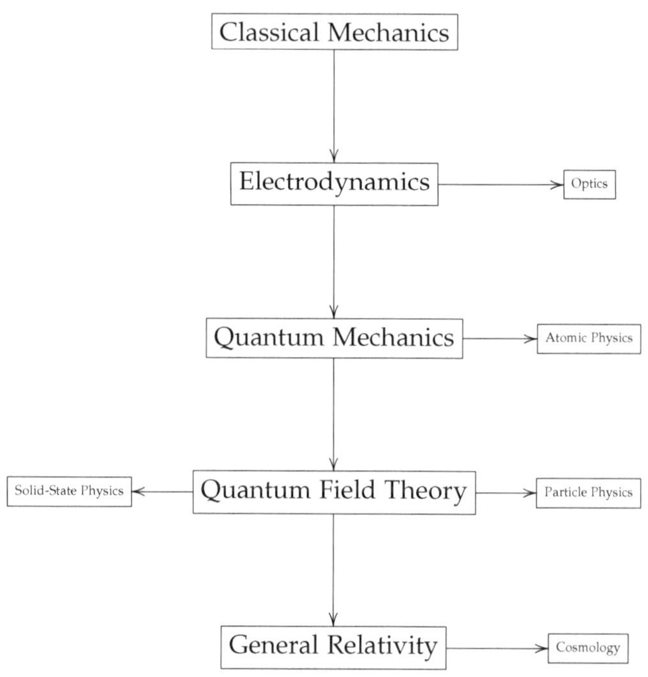

In contrast, if you follow a top-down approach you start with the most fundamental theories — quantum field theory, general relativity, statistical mechanics — and then study how the older theories emerge in appropriate limits. This approach is attractive because if you can't wait to study cool stuff like general relativity and quantum field theory, there is no need to spend months solving complicated classical mechanics problems first. So, whenever you don't feel like studying, you should start with the subject that interests you the most. Starting is often the most difficult part. Once you have started, it's much easier to continue. If I had to start again today, I would probably try to follow a top-down approach.

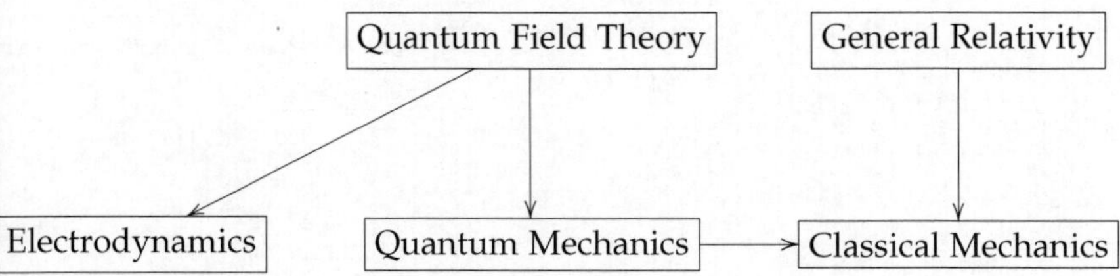

In any case, the fundamentals of each subject can often be grasped in a few weeks, and afterwards, you can decide if you want to dive deeper or move on to the next subject.

Irrespective of which route you follow, you should pick one excellent book for each subject to start with and then read it from cover to cover. I will recommend some of my favorites below. Of course, these books are just my favorites and certainly not a perfect fit for everyone. I'm only listing them below to give you a place to start. You can find many great alternatives by browsing Goodreads and Amazon reviews and by reading threads in online forums.

After finishing one first book on a given subject, you can decide whether you want to move on or dive deeper using additional resources. It should be clear that no single book can make you an instant expert. It is simply impossible to learn and understand everything about a given subject from a single book. But the books I recommend below will give you an extremely solid foundation and allow you to understand the fundamentals. Afterwards, you will find it straightforward to build up from there.

We will talk about these matters in more detail in Part V. But first, let's just dive right in and talk about the fundamental theories of physics.

7

Classical Mechanics

At A Glance

Why study it?

It's still the best theory of everyday objects that we have. And classical mechanics is extremely helpful if we want to understand the various mathematical arenas and how we can describe a given theory using different formulations.

Recommended book:

The Lazy Universe by Jennifer Coopersmith

Main learning objectives:

To thoroughly understand fundamental concepts like configuration space, phase space and the Lagrangian formalism as they are used everywhere in modern physics.

7.1 Why Learn Classical Mechanics?

The great thing about classical mechanics is that we can use it to describe everyday objects. The behavior of these objects is completely in agreement with our everyday experiences and nothing strange happens. While this may sound boring, it is actually exciting news.

The ordinary content of what we describe using classical mechanics allows us to focus on the tools that we use. More specifically, we can learn how to use a physical theory to describe a concrete system. In addition, we learn fundamentals concepts like what an equation of motion is and how we can solve it.

And we can also use classical mechanics to become familiar with many somewhat abstract concepts (e.g. Noether's theorem, configuration space, phase space, changes of the coordinate system, the Hamiltonian and the Lagrangian), that are used everywhere in modern physics.

It's much harder to learn these fundamentals concepts while trying to describe, say, quantum systems. Understanding all the unintuitive things that happen in such systems is difficult enough, even if you're already familiar with all the necessary tools.

And finally, classical mechanics, of course, has a lot of practical value. Ideas like the principle of virtual work are essential for engineers who want to make sure that buildings are stable and without classical mechanics, no rocket would be able to land on Mars.

Now, before we talk about how exactly you should learn classical mechanics, let's discuss briefly what it's all about.

7.2 Bird's Eye View of Classical Mechanics

The Newtonian formulation is the standard formulation of classical mechanics that every kid learns in high school. The main idea is that we can describe how the various objects in our system interact with each other in terms of forces. We can then calculate how a given object will move by collecting all forces that act on it and putting this sum into Newton's second law:

$$\underbrace{\frac{d}{dt}\vec{p}}_{\text{rate of change of the momentum}} = \overbrace{\vec{F}}^{\text{total force}} \tag{7.1}$$

This way, we get the equation of motion for the object in question. The solution of this equation for fixed initial conditions describes the path of the object. The function that describes the path of a given object is usually called a trajectory.

A much more magical way to describe classical mechanics is the Lagrangian formulation. In classical mechanics, the Lagrangian is simply the kinetic energy minus the potential energy:

$$\underbrace{L}_{\text{Lagrangian}} = \overbrace{T}^{\text{kinetic energy}} - \underbrace{V}_{\text{potential energy}} \tag{7.2}$$

Using the Lagrangian we can define the action, which is

what we get if we integrate the Lagrangian over time:

$$\underbrace{S[q(t)]}_{\text{action}} \equiv \overbrace{\int_{t_i}^{t_f} dt}^{\text{time integral}} \underbrace{L\Big(q(t), \dot{q}(t), t\Big)}_{\text{Lagrangian}} \tag{7.3}$$

The main idea is that any object will always follow the path of least action.

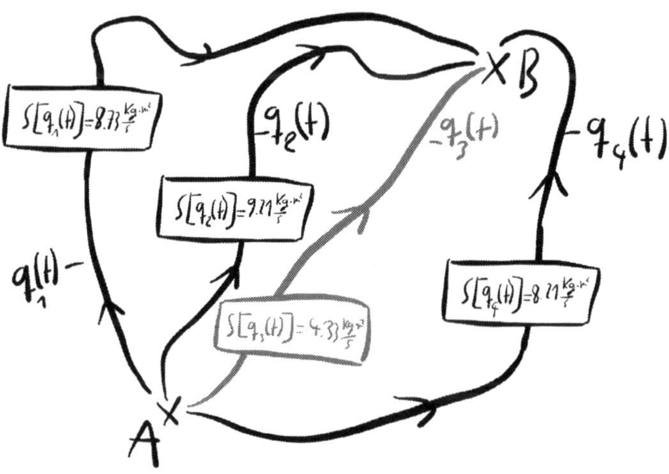

This means, at each possible location, we can imagine that each further step the object moves corresponds to a specific amount of action. The total action of a given path is the sum of these individual contributions.

The magical thing is that nature really always chooses the path which requires the minimum amount of total action.

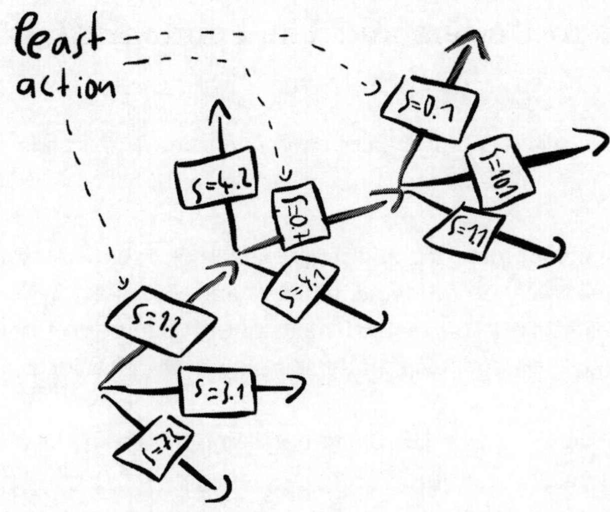

Therefore, the main idea behind the Lagrangian formulation is that nature is lazy!

A third way to describe classical mechanics is the Hamiltonian formalism. Here, our main formulas are called Hamilton's equations. In some sense they are a way of rewriting Newton's second law of motion as a first-order system of differential equations (i.e, in such a way that they contain only first order time derivatives). The price we have to pay for this simplification is that we double our degrees of freedom since we treat the momenta of our objects as independent variables.

7.3 How To Understand The Fundamentals

The best classical mechanics book I've come across so far is The Lazy Universe by Jennifer Coopersmith[1].

It's really an inspiring and passionate book. Coopersmith focuses solely on fundamental ideas and discusses them all in great detail. This makes The Lazy Universe perfect if you want to develop a deep understanding of the core ideas.

While most other textbook authors simply state the facts and convey the attitude: "It works, so stop asking why", Coopersmith always discusses at great length why a specific concept is introduced. That way, she makes sure that the reader always stays interested, which is especially important for self-learners.

You should consult additional good introductory textbooks whenever you feel like you didn't fully understand what Coopersmith is saying or to see more real world applications. I recommend "Classical Mechanics" by John R. Taylor, "Introduction to Classical Mechanics" by David Morin and "Classical Dynamics" by David Tong[2]. In addition, there is a wonderful little book called "The Variational Principles of Mechanics" by Cornelius Lanczos which was the main inspiration for Coopersmith's book. And finally, you might enjoy my book "No-Nonsense Classical Mechanics"[3].

You should also be aware that the standard text, which is commonly used for lectures, is Classical Mechanics by Herbert Goldstein.

[1] Jennifer Coopersmith. *The lazy universe : an introduction to the principle of least action*. Oxford University Press, Oxford New York, NY, 2017. ISBN 978-0-19-874304-0)

[2] `http://www.damtp.cam.ac.uk/user/tong/dynamics.html`

[3] Jakob Schwichtenberg. *No-Nonsense Classical Mechanics : a student-friendly introduction*. No-Nonsense Books, Karlsruhe, Germany, 2019b. ISBN 9781096195382

Key concepts to focus on:

▷ Newton's second law and how to solve the resulting equation of motion for specific systems.

▷ The Lagrangian function and the action functional.

▷ The basic idea at the heart of variational calculus and how we can use it to find minima of the action.

▷ The Euler-Lagrange equation.

▷ Noether's theorem.

▷ Hamilton's equations.

▷ Poisson brackets.

Example questions:

▷ What's the connection between the Lagrangian function and the Hamiltonian function?

▷ In what sense do Hamilton's equations follow directly from the fact that partial derivatives commute?

▷ What's the equivalent to Noether's theorem in the Hamiltonian formalism?

▷ In what sense do Poisson brackets describe the time evolution of observables?

7.4 Map of Classical Mechanics

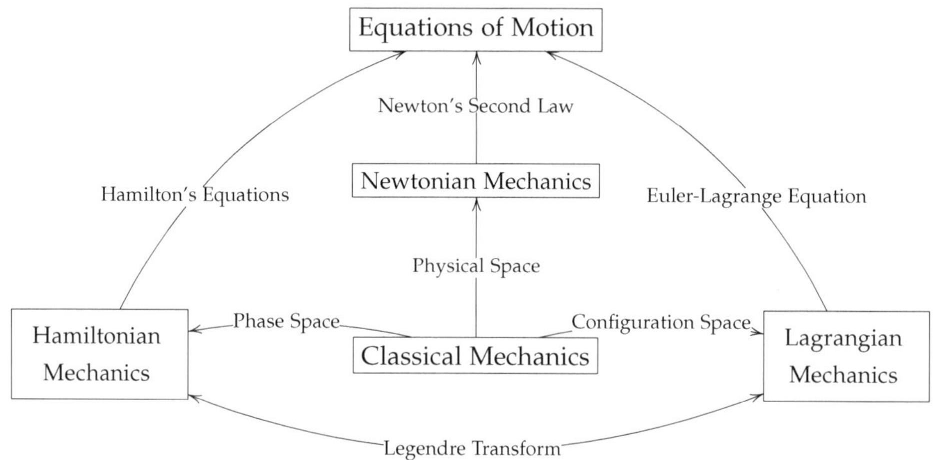

7.5 How To Level Up

The best way to get a deep understanding of the main ideas of classical mechanics and really grasp the connections between the various formulations is to use geometry. It can be extremely useful to learn about abstract concepts like symplectic geometry and fiber bundles in the relatively simple context of classical mechanics. Fiber bundles, especially, are a unifying theme in all modern theories of physics.

A great starting point to learn more about the geometric interpretation of classical mechanics are the Lectures on "Classical Mechanics" by John C. Baez and Derek K. Wise[4]. The standard reference, which is quite accessible, is "Mathematical Methods of Classical Mechanics" by Vladimir I. Arnold[5].

[4] http://math.ucr.edu/home/baez/classical/texfiles/2005/book/classical.pdf

[5] [Arnold, 1989]

And take note that there is a fourth formulation of classical mechanics, called the Koopman–von-Neumann formulation, which is rarely discussed in the literature. Nevertheless, it is interesting if you want to understand the connection between classical mechanics and quantum mechanics. A good starting point to learn more about it is the article "Koopman-von-Neumann Mechanics, and a Step Beyond" by Nobel laureate Frank Wilczek.[6]

[6] http://frankwilczek.com/2015/koopmanVonNeumann02.pdf

8

Electrodynamics

At A Glance

Why study it?
It's the prototypical example of a classical field model and is useful to understand all other models of fundamental interactions.

Recommended book:
A Student's Guide to Maxwell's Equations by Daniel A. Fleisch

Main learning objectives:
To understand the concepts "field" and "gauge symmetry" as deeply as possible. Both are at the heart of everything we know about nature at the most fundamental level.

8.1 Why Learn Electrodynamics?

First of all, electrodynamics correctly describes the behavior of one of only four known fundamental interactions.

At large (cosmological) scales, gravity is the most important of the four interactions.[1] At tiny scales, strong and weak interactions are responsible for most of the interesting phenomena. For example, protons are held together by the strong force and atoms decay through weak interactions. However, in between these two extremes electromagnetic interactions hold sway. Practically speaking, electrodynamics not only allows us to understand how electricity and magnetism work, but also what light is and how it travels.

[1] This is true at least on a macroscopic level. Electromagnetic interactions of elementary particles are described by quantum electrodynamics which is a model within the framework of quantum field theory.

Plus, a solid understanding of electrodynamics is useful to understand the theories describing all other fundamental interactions. And the equations describing electromagnetic interactions of macroscopic objects (Maxwell's equations) are exactly the same as the equations governing electromagnetic interactions of elementary particles. The equations describing other fundamental interactions are extremely similar and can be understood much easier once the equations of electrodynamics are understood. This means, the knowledge gathered by studying electromagnetic interactions on a macroscopic level helps us immediately to understand what happens on a more fundamental level. Therefore, electrodynamics is an ideal playground to understand many of the most important concepts underlying modern physics in a simplified setup. In the context of electrodynamics we can understand the following:

▷ What gauge symmetry is

▷ What a gauge field is. How we can interpret it geometrically. And why gauge fields are responsible for fundamental interactions

▷ What special relativity is all about

Lastly, electrodynamics is prototypical for what progress in physics means. Before it was developed, Leonhard Euler made the following statement:[2]

[2] Quoted in [Zangwill, 2013].

> "The subject I am going to recommend to your attention almost terrifies me. The variety it presents is immense, and the enumeration of facts serves to confound rather than to inform. The subject I mean is electricity."

Nowadays, all we have to do to understand the multitude of electric and magnetic phenomena is study five short equations and you don't have to be a genius to master electricity. Framed differently, Maxwell's model of electrodynamics is certainly one of the most important scientific works of all time. This is especially true if we use David Hilbert's criterion:[3]

[3] Quoted in [Eves, 2003].

> "One can measure the importance of a scientific work by the number of earlier publications rendered superfluous by it."

Before Maxwell developed his model of electrodynamics, there were hundreds of publications, each describing a different electric and magnetic phenomenon, or properties of light. These days, we only need Maxwell's equations to describe all this.

Next, let's discuss what Maxwell's theory of electrodynamics is all about.

8.2 Bird's Eye View of Electrodynamics

Our goal in electrodynamics is to describe the interplay between electrically charged objects and the electromagnetic field. A field is a mathematical object that eats a location and point in time and returns a specific other mathematical object. For example, a vector field eats a location and point in time (i.e. a spacetime point) and spits out a vector.

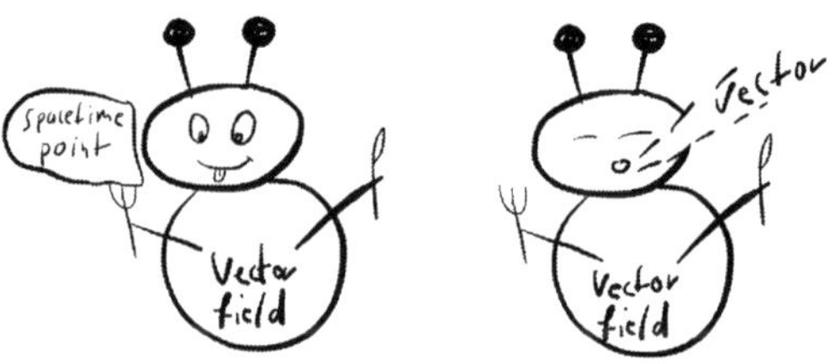

Since it does this for every location and point in time (i.e. every spacetime point), the picture emerges that a vector field is like a collection of arrows that float above spacetime.

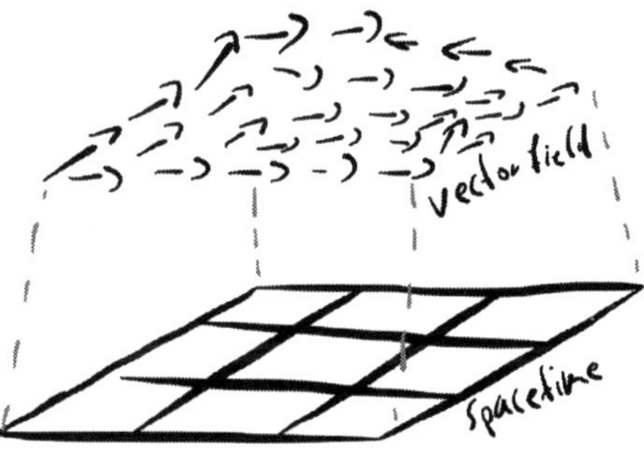

Electrically charged objects generate nontrivial structures in the electromagnetic field. This is described by Maxwell's equations. In turn, nontrivial structures in the electromagnetic field have a direct influence on electrically charged objects and this is described by the Lorentz force law.

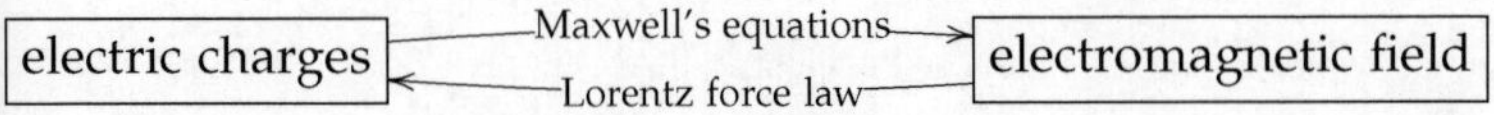

electric charges —Maxwell's equations→ electromagnetic field
electric charges ←Lorentz force law— electromagnetic field

Maxwell's equations are really the heart of electrodynamics. The first one is known as Gauss's law for the electric field and tells us that an electric charge produces a radial electric field:

divergence of the electric field

$$\nabla \cdot \vec{E} = \frac{\rho}{\epsilon_0}. \tag{8.1}$$

charge density divided by the vacuum permittivity

In general, the divergence of a vector field, $\nabla \cdot \vec{F}$ encodes whether or the field lines point radially away from a given location:

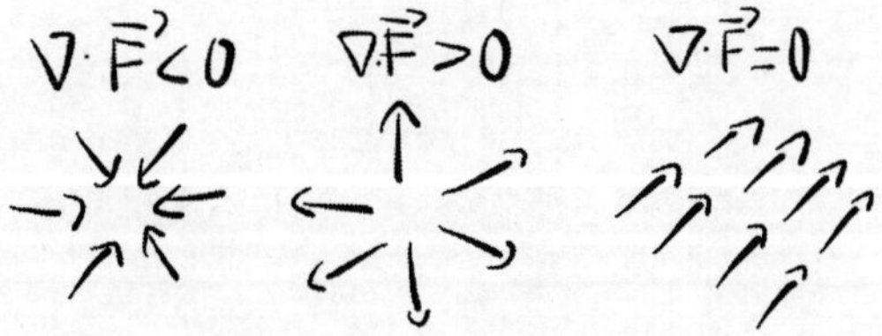

$\nabla \cdot \vec{F} < 0$ $\nabla \vec{F} > 0$ $\nabla \cdot \vec{F} = 0$

The second one is known as Gauss's law for the magnetic field and tells us that there are no magnetic charges (no magnetic monopoles), and therefore, no radial structures in the magnetic field:

divergence of the magnetic field

$$\nabla \cdot \vec{B} = 0. \tag{8.2}$$

The third one is known as Faraday's law and tells us that a changing magnetic field produces a circulating electric field:

$$\underbrace{\nabla \times \vec{E}}_{\text{curl of the electric field}} = -\underbrace{\frac{\partial}{\partial t}\vec{B}}_{\text{rate of change of the magnetic field}} \tag{8.3}$$

In general, the curl of a vector field, $\nabla \times \vec{F}$, encodes if there are circulating structures in it:

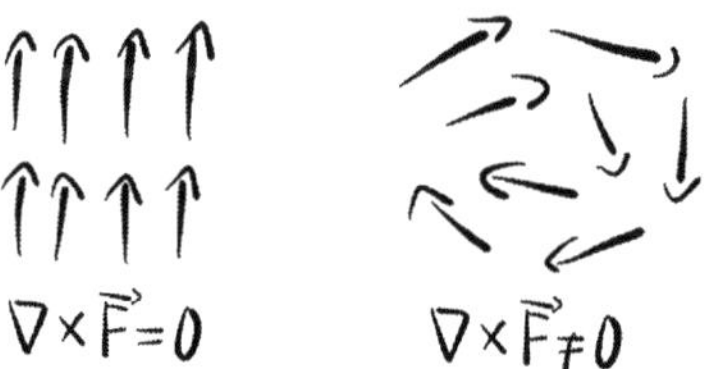

The final Maxwell equation is known as the Ampere-Maxwell law and tells us that an electric current and a changing electric field produce a circulating magnetic field.

$$\underbrace{\nabla \times \vec{B}}_{\text{curl of the magnetic field}} = \underbrace{\mu_0}_{\text{magnetic permeability of free space}} \Big(\underbrace{\vec{J}}_{\text{electric current}} + \underbrace{\epsilon_0}_{\text{electric vacuum permittivity}} \underbrace{\frac{\partial}{\partial t}\vec{E}}_{\text{rate of change of the electric field}} \Big). \tag{8.4}$$

These equations allow us to calculate what the electric and magnetic field look like in a given system.

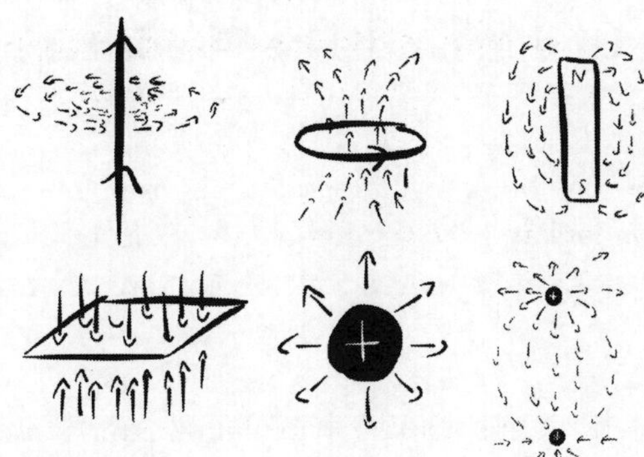

Moreover, using the Lorentz force law, we can then calculate how these electric and magnetic field structures influence charges that are present in the system:

$$\underbrace{\frac{d\vec{p}}{dt}}_{\text{rate of change of the momentum}} = \underbrace{q}_{\text{electric charge}}\Big(\underbrace{\vec{E}}_{\text{electric field}} + \underbrace{\frac{d\vec{x}}{dt}}_{\text{rate of change of the location}} \underbrace{\times}_{\text{cross product}} \underbrace{\vec{B}}_{\text{magnetic field}}\Big). \tag{8.5}$$

For example, we can use Lorentz force law to calculate how a charge gets deflected in the electric field that is generated by a capacitor.

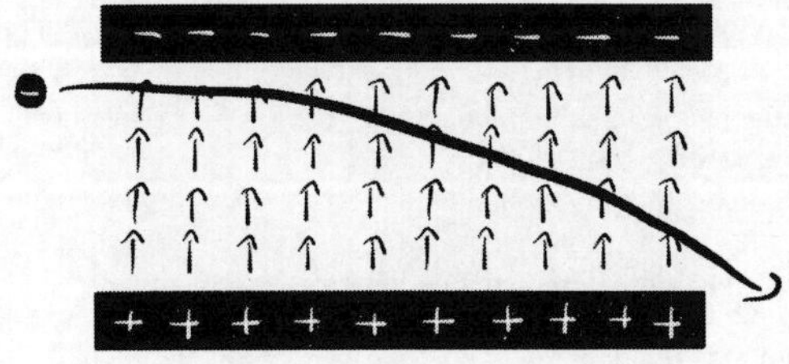

8.3 How To Understand The Fundamentals

My number one book recommendation for anyone who wants to understand the fundamentals of electrodynamics is "A Student's Guide to Maxwell's Equations" by Daniel Fleisch[4].

[4] Daniel Fleisch. *A student's guide to Maxwell's equations*. Cambridge University Press, Cambridge, UK New York, 2008. ISBN 978-0-511-38900-9

The sole focus of this book is to explain the fundamental concepts and equations used in electrodynamics in a student-friendly way. While this necessarily means that no real-world applications are discussed, it's the perfect book if you want to develop a solid understanding of the fundamentals.

Additional good introductory textbooks are the second volume of the "Feynman Lectures"[5] and the "Introduction to Electrodynamics" by David J. Griffith[6]. And again, you might enjoy my book "No-Nonsense Electrodynamics"[7].

[5] http://www.feynmanlectures.caltech.edu/

[6] David Griffiths. *Introduction to electrodynamics*. Cambridge University Press, Cambridge, United Kingdom New York, NY, 2017. ISBN 9781108420419

The standard reference which professors love to recommend, is "Classical Electrodynamics" by John D. Jackson.[8] Be warned, however, that Jackson's book is not really a good choice for beginners because he spends too much time on complicated calculations. And it's often not clear what is fundamental and what is only important for a certain application.

[7] Jakob Schwichtenberg. *No-Nonsense Electrodynamics*. No-Nonsense Books, Karlsruhe, Germany, 2019c. ISBN 9781790842117

[8] John Jackson. *Classical electrodynamics*. Wiley, New York, 1999. ISBN 9780471309321

Key concepts to focus on:

- ▷ Electric charge and charge density
- ▷ Current density
- ▷ The electric field and the magnetic field
- ▷ The flux and circulation of the electric and magnetic field
- ▷ The continuity equation

- ▷ The Lorentz force law
- ▷ Maxwell's equation and how we can use them to describe concrete system
- ▷ The wave equations and the basic properties of electromagnetic waves

Example questions:

- ▷ In what sense do the homogeneous Maxwell equations follow directly from the geometrical statement that the boundary of a boundary is always zero? (Hint: Bianchi identities).
- ▷ How exactly can we see that electric charge is locally conserved and not only globally?
- ▷ What is more fundamental: the electric field, the magnetic field, the electromagnetic field tensor, or the electromagnetic potential?
- ▷ How can special relativity be "derived" from the equations of electrodynamics?
- ▷ Can the electromagnetic potential be measured?

8.4 Map of Electrodynamics

Formalism:

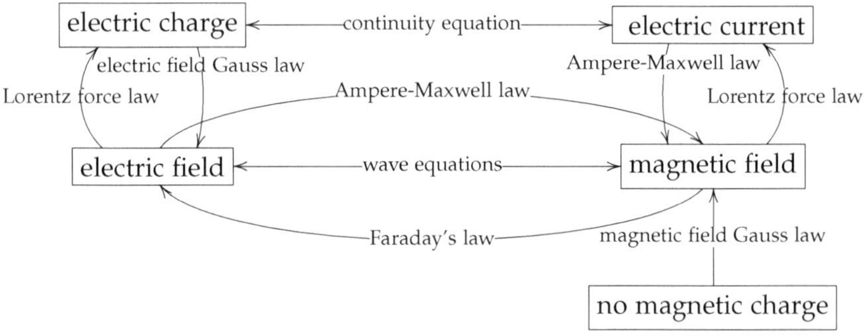

Ingredients

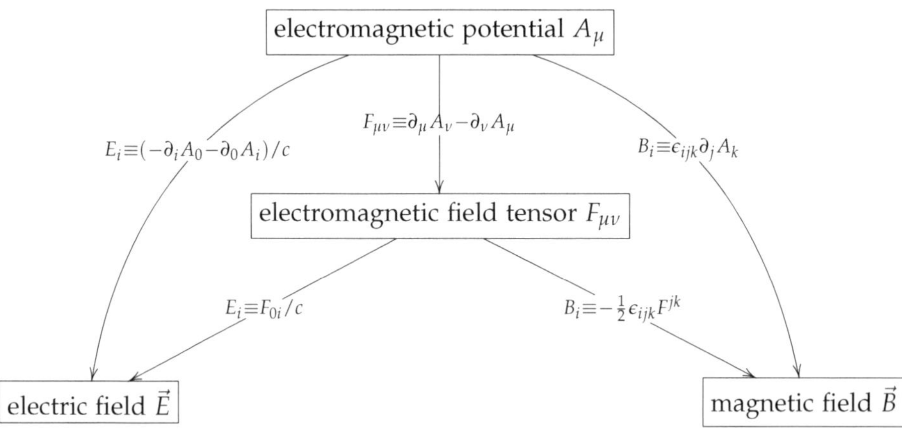

8.5 How To Level Up

There are several possibilities to go beyond the usual description of electrodynamics and develop a deeper understanding. Two important ones are directly connected to Einstein's theory of special relativity, one to how we can describe all fundamental interactions in a common context, and another one to how we can describe electrodynamics in geometric terms.

Let's discuss these ideas one by one.

One of the basic facts described by electrodynamics is that a static charge leads to a nonzero electric field but has no effect on the magnetic field. On the other hand, a moving charge leads to both, a nonzero electric field and a nonzero magnetic field.

But these simple facts can quickly lead to a lot of confusion. For example, let's imagine that we observe a static charge and therefore conclude that there is no resulting magnetic field ($\vec{B} = 0$). However, a second observer who moves relative to us describes the charge as a moving charge and therefore concludes that there is a nonzero magnetic field ($\vec{B} \neq 0$). Who is right?

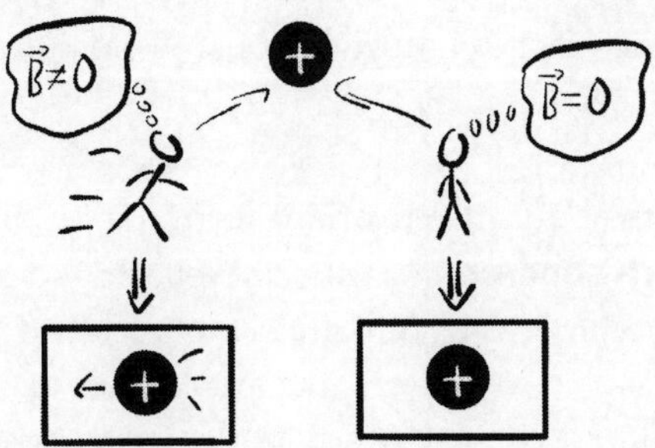

It turns out that, of course, there is no paradox here. Instead, what we learn here is simply that the distinction between the electric and magnetic field depends on how we observe the field. There is really only one field called the electromagnetic field and for some observers it appears solely like an electric field while for other observers it appears as an electric field and a magnetic field simultaneously. In other words, the electric field and magnetic field are part of a larger structure called the electromagnetic field. This is analogous to how we learn in special relativity that we can't really think of space and time as two separate concepts. Instead, both are part of a larger structure we call spacetime.

To describe all of this consistently, we need to rewrite the laws of electrodynamics in terms of this unifying field. The resulting equations are known as the covariant form of Maxwell's equations. The most important keywords in this context are electromagnetic potential and electromagnetic field tensor.

A second important observation is that the waves described by the vacuum wave equations always move with a velocity of exactly 299,792,458 m/s. This is a strange result because, usually, the velocity of objects depends on how we measure it. For example, a person who runs parallel to a train measures a different velocity than an observer who stands still. For electromagnetic waves this is not the case. Taking this curious fact of nature seriously leads us directly to Einstein's theory of special relativity.

To learn more about the covariant formulation of electrodynamics and its connection to special relativity try Chapter 12 in Introduction to Electrodynamics by David J. Griffith[9].

[9] David Griffiths. *Introduction to electrodynamics.* Cambridge University Press, Cambridge, United Kingdom New York, NY, 2017. ISBN 9781108420419

Another important aspect of electrodynamics worth focusing on is gauge symmetry and in what sense we can understand electrodynamics as a gauge model. This is one of the most important insights in all of physics because all of our best models of fundamental interactions are gauge models. And electrodynamics is the simplest context to develop some understanding of the most important ideas that are commonly used in gauge models. However, be warned that this is a somewhat advanced topic usually only discussed properly in the context of quantum field theory.[10]

[10] Nevertheless, I try to explain the main ideas in a way that is hopefully understandable for undergraduates in Chapter 8 of my book No-Nonsense Electrodynamics.

In a similar spirit, it makes sense to study how we can understand electrodynamics geometrically. This perspective reveals the deep connection to all other models of fundamental interactions. And it allows us to understand the deep meaning of the electromagnetic potential and the electromagnetic field tensor. Studying electrodynamics in geometrical terms is possibly using a branch of mathematics known as differential geometry. For a gentle introduction try "A pictorial introduction to differential geometry, leading to Maxwell's equations as three pictures" by Jonathan Gratus[11].

[11] https://arxiv.org/abs/1709.08492

9

Special Relativity

At A Glance

Why study it?
All fundamental theories of physics must obey the laws of special relativity.

Recommended book:
Special Relativity by Anthony P. French

Main learning objectives:
To understand the wide-ranging implications of the simple but curious fact of nature that the speed of light (in a vacuum) always has the same value no matter how you move relative to it.

9.1 Why Learn Special Relativity?

"All the laws of physics are constrained by special relativity acting as a sort of "super law". " **Jean-Marc Levy-Leblond**[1]

[1] Jean-Marc Lévy-Leblond. One more derivation of the Lorentz transformation. *American Journal of Physics*, 44(3):271–277, Mar 1976. DOI: 10.1119/1.10490

Anyone interested in how nature works should have a basic understanding of special relativity because the best theory of elementary particles that we have (quantum field theory) is what we end up with if we combine the lessons of quantum mechanics, with the lessons of special relativity. Moreover, even in general relativity the main ideas of special relativity remain relevant. So, in other words, the main ideas of special relativity are at the very heart of our modern understanding of how nature works.

While special relativity reveals various, weird features of nature, they all proved to be correct when tested in experiments. This is an important point because oftentimes the results we get using special relativity imply things very different from our everyday experiences. But nature doesn't care about what we deem strange. As long as the things we predict are confirmed experimentally, everything is alright.

Last but not least, from a more practical perspective, modern navigation systems like GPS wouldn't be possible without knowledge of special relativity. So, special relativity also has important technological implications. This is especially true if you want to think about interplanetary space travel.

Now, before we talk about how you should learn special relativity in concrete terms, let's discuss briefly what it is all about.

9.2 Bird's Eye View of Special Relativity

These are the key ideas at the heart of special relativity:

1. The laws of physics are the same for all (inertial) observers.
2. The speed of light in a vacuum has the same value for all (inertial) observers.

While the first idea seems reasonable, the second one is really crazy. To understand why, let's do a small thought experiment.

Let's imagine that we measure the speed of a train twice. The first time we simply stand still. The second time we measure the train's speed while running next to it. The main observation here is that we will measure a lower (relative) velocity if we run. For example, if we measure 50 km/h while standing, we will measure 35 km/hr if we run with a speed of 15 km/hr parallel to the train. But if we do the same thing with an electromagnetic wave (light), we find something completely different! We always measure exactly the same value no matter how we move.

There are two ways to end up with this conclusion. The first one is the famous Michelson-Morley experiment that historically discovered this crazy property of electromagnetic waves. A second possibility is to have a sharp look at Maxwell's equations of electrodynamics.

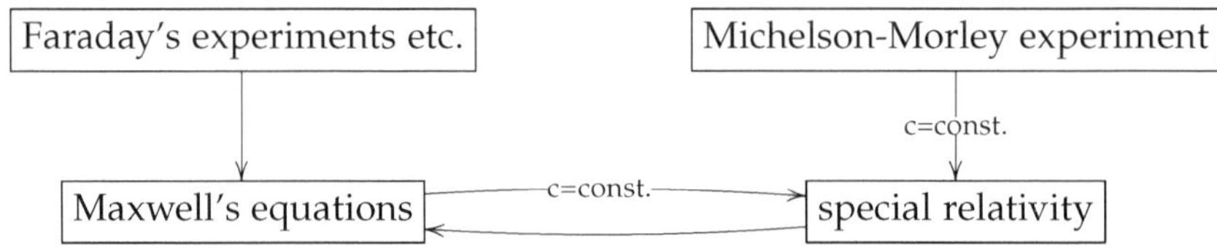

What we end up with, if we take the constancy of the speed of light seriously, is a model that accurately predicts how the basic measurable quantities of physics (time, distance, mass, and energy) depend on the relative speed of the measuring apparatus.

The equations of special relativity tell us how these quantities must change to guarantee that the value of the speed of light remains fixed and the laws of physics for all experimenters are exactly the same. For example, an observer moving fast relative to a stick measures a shorter length than an observer moving slowly. Similarly, two observers moving with different constant velocities do not agree on the number of seconds that have passed between two given events.

To sum up, depending on how a given observer moves, a little bit of space will look like a little bit of time or vice versa.

To describe this interplay, we end up with the notion of spacetime that combines the spatial dimensions and the time axis into a single object:

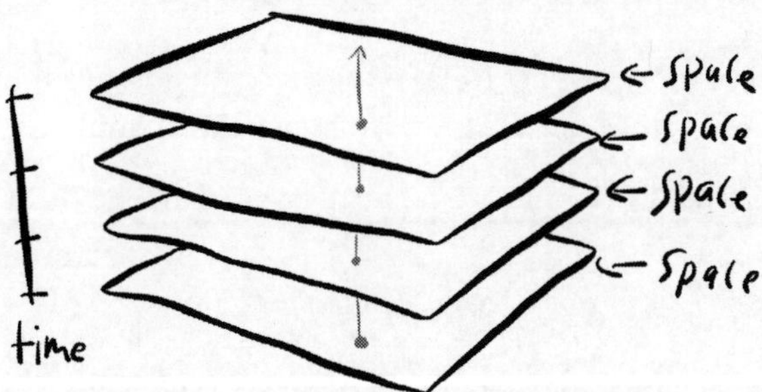

The constancy of the speed of light is hard coded in the structure of spacetime. To understand how, take note that a point in spacetime corresponds to an "event". For example, if an object is located at some specific location at a specific instance in time we call this an event. In mathematical terms, we describe each event by a four vector

$$A = \begin{pmatrix} x_0 \\ x_1 \\ x_2 \\ x_3 \end{pmatrix}, \tag{9.1}$$

where the zeroth component encodes the corresponding instance in time and the remaining three components the event's spatial location.

An important question is then: what's the "distance" between two events in spacetime? Naively, we would think that the distance between an event A and another event

$$B = \begin{pmatrix} \tilde{x}_0 \\ \tilde{x}_1 \\ \tilde{x}_2 \\ \tilde{x}_3 \end{pmatrix} \tag{9.2}$$

is given by

$$d(A,B) = \Delta x_0^2 + \Delta x_1^2 + \Delta x_2^2 + \Delta x_3^2 \,. \tag{9.3}$$

But this is not correct. To take the fact into account that the speed of light is constant, we must use the formula

$$\begin{aligned} d(A,B) &= \Delta x_0^2 - \Delta x_1^2 - \Delta x_2^2 - \Delta x_3^2 \qquad \text{↷ } \Delta x_0 \equiv c\Delta t \\ &= (c\Delta t)^2 - \Delta x_1^2 - \Delta x_2^2 - \Delta x_3^2 \,, \end{aligned} \tag{9.4}$$

where c denotes the speed of light. Mathematically, this implies that we consider events in "Minkowski space" instead of events in "Euclidean space".

9.3 How To Understand The Fundamentals

A really great book to understand the fundamentals of special relativity is Spacetime Physics by Edwin F. Taylor and John Archibald Wheeler[2]. This is one of the rare exceptions where two leading experts have written an extremely student-friendly book.

[2] Edwin Taylor and John Wheeler. *Spacetime physics : introduction to special relativity*. W.H. Freeman, New York, 1992. ISBN 9780716723271

Wheeler and Taylor manage to explain all fundamental aspects of special relativity as gently as possible using lots of nice illustrations. Another great aspect is that the book is available for free on the author's website.[3]

[3] http://www.eftaylor.com/

9.4 Map of Special Relativity

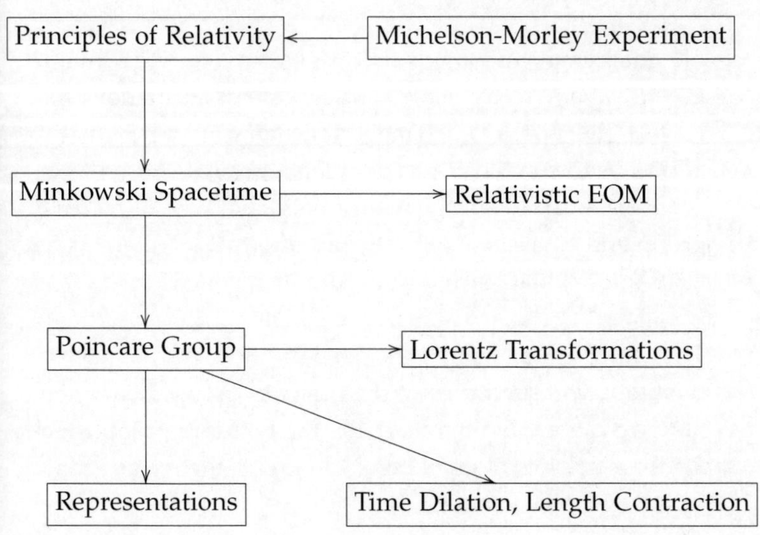

9.5 How To Level Up

The key insight to understand why special relativity is so important in modern physics is to recognize that the constancy of the speed of light reveals a fundamental symmetry of nature. In general, a symmetry is a set of transformations that keep the object in question unchanged. Using special relativity, we learn that the speed of light must remain unchanged, and so, we can start to think about all allowed transformations for which this is the case. The set of these transformations is known as the Poincare symmetry and lies at the heart of the Standard Model of Particle Physics and general relativity.

Knowing that our laws must be invariant under all Poincare transformations is an extremely powerful restriction and is almost sufficient to derive all fundamental equations (Klein-Gordon equation, Maxwell's equations, and the Dirac equation).

The mathematical toolbox that we use to describe symmetries is called group theory. The set of transformations that leave a given object unchanged together with a rule that allows us to combine them is called a group. The symmetry at the heart of special relativity is called the Poincare group and consists of all transformations (e.g. rotations and translations) that leave the speed of light unchanged.

Therefore, to understand the deep implications of special relativity, you should learn the basics of group theory and how we can apply them to study the Poincare group. Luckily, there are lots of great books on group theory, as Predrag Cvitanovic once pointed out:

> "Almost anybody whose research requires sustained use of group theory (and it is hard to think of a physical or mathematical problem that is wholly devoid of symmetry) writes a book about it."

And case in point, I've written one myself titled "Physics from Symmetry".[4] But there are really great alternatives like "Symmetries in Fundamental Physics" by Kurt Sundermeyer[5] and "Symmetry and the Standard Model" by Matthew Robinson[6]. Be warned, however, that all these books go far beyond special relativity because it's really hard to not get excited about all the cool things we can learn once we understand group theory.

[4] Jakob Schwichtenberg. *Physics from symmetry*. Springer, Cham, Switzerland, 2018a. ISBN 978-3-319-66631-0

[5] Kurt Sundermeyer. *Symmetries in fundamental physics*. Springer, Cham, 2014. ISBN 9783319065809

[6] Matthew Robinson. *Symmetry and the standard model : mathematics and particle physics*. Springer, New York, 2011. ISBN 9781441982667

10

Quantum Mechanics

At A Glance

Why study it?
It's the crucial stepping stone towards quantum field theory. Without understanding all relevant notions in the simplified setup of quantum mechanics it's much harder to use them in quantum field theory.
Recommended book:
Quantum Mechanics by David J. Griffith
Main learning objectives:
To learn the standard quantum formalism, i.e. how we can describe what happens in a given system using objects in Hilbert space.

10.1 Why Learn Quantum Mechanics?

The strongest argument in favor of quantum mechanics is that experiments tell us that it is correct. Physicists and philosophers can argue all day long about how beautiful some theory is, but in the end, the only thing that matters is if it agrees with what we observe in experiments. And quantum mechanics certainly does. Similar to how you need classical mechanics to describe how a ball rolls down a ramp, we need quantum mechanics to describe something like the hydrogen atom

In addition, quantum mechanics is the little brother of quantum field theory, which is the best theory of the fundamental interactions that we have. And it is quite hard to understand many important aspects of quantum field theory without learning the basic concepts in the simpler context of quantum mechanics first.

To summarize: quantum mechanics is an essential tool in the repertoire of any competent physicist. While it will not help you to describe the ball that rolls down a ramp any better, it will allow you to describe new physical systems that you can't describe with classical mechanics.

Now, before we talk about how you should learn quantum mechanics in concrete terms, it makes sense to discuss briefly what it is all about.

10.2 Bird's Eye View of Quantum Mechanics

The laws governing quantum systems are very different from what we would expect from our everyday experiences.

The observation at the heart of quantum mechanics is that we need wave functions to describe particles like an electron, for example. A wave function is a mathematical object that behaves, well, like a wave. This is confusing because a wave is something that spreads out in space while a particle is strictly localized at a single point. So, unsurprisingly, no theoretical physicist predicted this interplay between waves and particles. Instead, as is usual in physics, we were forced to accept this idea through experiments. The most famous experiment that demonstrated that we need wave functions to describe particles, is the double-slit experiment. Waves and particles both produce a certain pattern if we shoot them onto a double slit. However, if we shoot electrons onto a double slit, we get the pattern that we usually associate with waves.

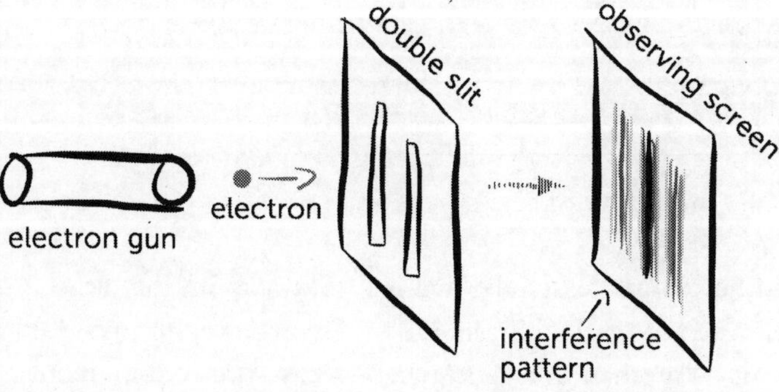

But electrons are certainly particles because the detector

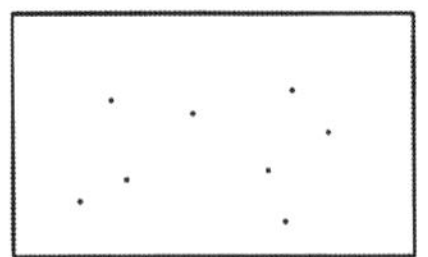

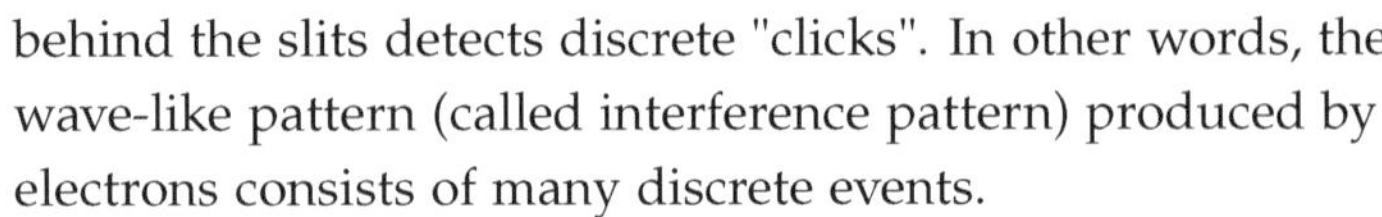

behind the slits detects discrete "clicks". In other words, the wave-like pattern (called interference pattern) produced by electrons consists of many discrete events.

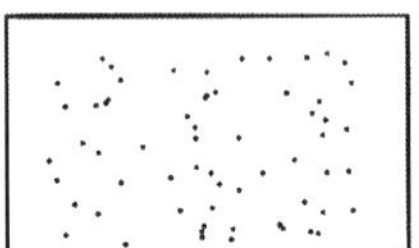

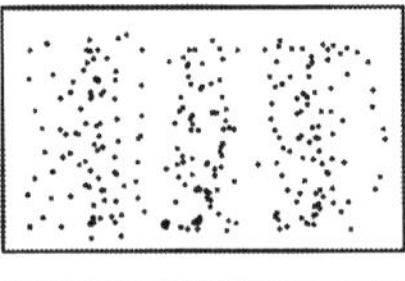

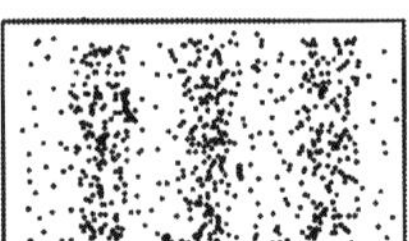

One direct implication of our need for a wave description is that we always have to deal with uncertainty. This is the case because if we describe particles using waves, certain properties of the wave have to correspond to particle properties. For example, the "location" of the wave tells us where the particle is, and its wave length is directly connected to the momentum of the particle. Now, for a wave that is clearly localized within some region, it's easy to understand how it describes the location of the corresponding particle:

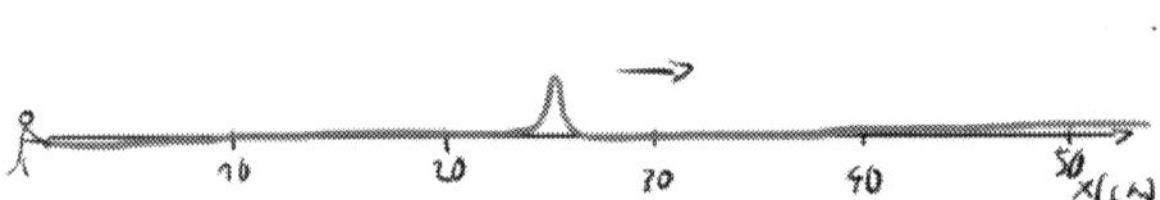

But what's the wave length here? There is no good answer and, therefore, we are uncertain about the momentum of the particle described by such a wave. Similarly, if we are dealing with a wave that has a perfectly obvious wavelength, there is no good way to say where exactly the wave is.

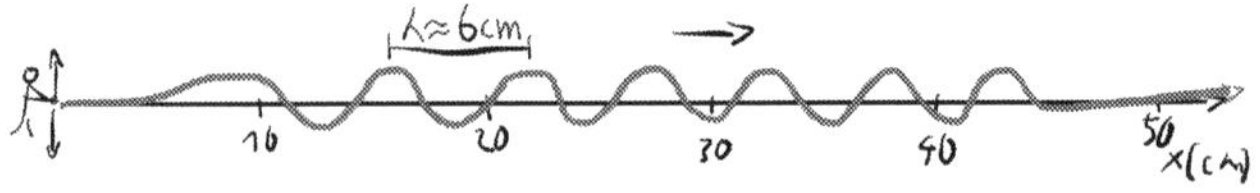

In other words, whenever we are certain about the location of the particle described by the wave, we are uncertain about its momentum. And whenever we are certain about the momentum of the particle, we are uncertain about its location. This fundamental uncertainty is one of the key features of quantum mechanics. Mathematically, it is directly related to the canonical commutation relation

$$\underbrace{[\hat{p}, \hat{x}]}_{\text{commutator of location and momentum}} \underbrace{\equiv}_{\text{is defined as}} \hat{p}\hat{x} - \hat{x}\hat{p} = -\underbrace{i}_{\text{imaginary unit}} \underbrace{\hbar}_{\text{reduced Planck constant}} \tag{10.1}$$

Here, the (reduced) Planck constant encodes the magnitude of quantum effects. In words, the canonical commutator tells us that it makes a difference whether we first measure the location of an object and then its momentum, or if we first measure its momentum and then its location.

To get an intuitive understanding for what it means for operators to have a non-zero commutator, compare the situation where you first put on your socks and then your shoes with the situation where you first put on your shoes and then your socks. The outcome is clearly different — the ordering of the operations "putting shoes on" and "putting socks on" therefore matters. In technical terms, we say these two operations do not commute. In contrast, it makes no difference if you put on your left sock first and then your right sock or your right sock first and then your left sock. These two operations do commute.

Another interesting thing we can derive as soon as we have accepted that we describe particles using waves is that certain fundamental quantities become quantized. This means, for example, that in some quantum systems only a discrete set of energy levels is allowed. We can understand this by noting that if we are dealing with a system with fixed boundaries (e.g. a box) only certain wave forms fit in. Each allowed wave form corresponds to a state with well-defined energy our particle can be in. Since only certain wave forms are allowed, we can conclude that only certain energy levels are allowed.

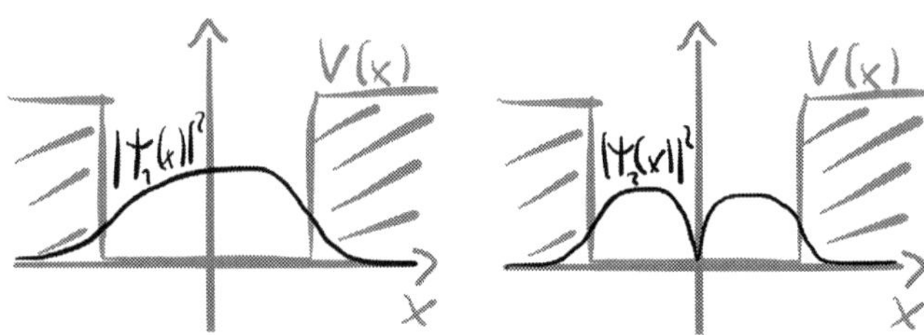

We find the wave function that describes a given quantum system by solving the Schrödinger equation:

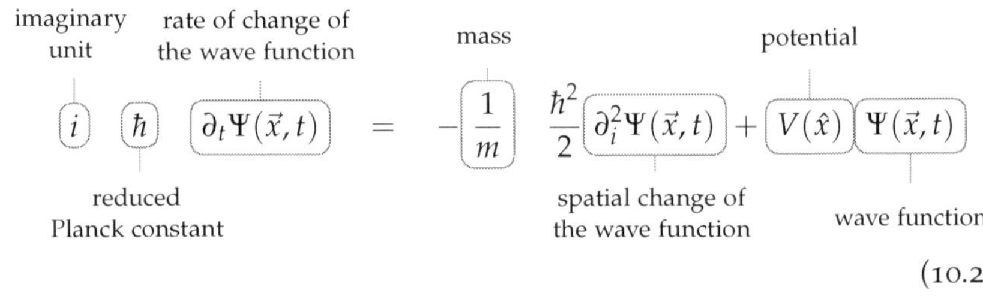

$$i \hbar \partial_t \Psi(\vec{x}, t) = -\frac{1}{m} \frac{\hbar^2}{2} \partial_i^2 \Psi(\vec{x}, t) + V(\hat{x}) \Psi(\vec{x}, t) \tag{10.2}$$

10.3 How To Understand The Fundamentals

My favorite quantum mechanics textbook is Quantum Mechanics by David J. Griffith[1].

[1] David Griffiths. *Introduction to quantum mechanics*. Pearson Prentice Hall, Upper Saddle River, NJ, 2005. ISBN 9780131118928

Griffith explains all fundamental aspects of quantum mechanics in great detail and always makes sure you understand how each concept is used in practice. And in contrast to most other books, he dives right in without 100+ pages of preparation chapters. This makes the book extremely enjoyable to read and ideal for self-study.

Other good introductory textbooks you can consult are the third volume of the Feynman Lectures on Physics[2] and Principles of Quantum Mechanics by Ramamurti Shankar[3]. In addition, you might also enjoy my book No-Nonsense Quantum Mechanics[4].

[2] `http://www.feynmanlectures.caltech.edu/`

[3] Ramamurti Shankar. *Principles of quantum mechanics*. Plenum Press, New York, 1994. ISBN 9780306447907

[4] Jakob Schwichtenberg. *No-nonsense quantum mechanics : a student-friendly introduction*. No-Nonsense Books, Karlsruhe, Germany, 2018b. ISBN 9781790455386

The standard reference that professors love to recommend is Quantum Mechanics by Claude Cohen-Tannoudji[5]. But it's not really an inviting introductory book because it starts with around 100 pages of math without any proper motivation, and is generally lacking insightful comments on what is really happening in calculations.

[5] Claude Cohen-Tannoudji. *Quantum mechanics*. Wiley, New York, 1977. ISBN 9780471164333

Key concepts to focus on

- ▷ Schrödinger equation
- ▷ Canonical commutation relation

- ▷ Quantum operators
- ▷ Uncertainty relations
- ▷ Eigenvalues and eigenstates
- ▷ Expectation value and standard deviation
- ▷ Bra-Ket notation
- ▷ Probability amplitudes
- ▷ Quantum harmonic oscillator and ladder operators
- ▷ Spin

Example questions

- ▷ In what sense is the usual wave function only a special case if we look at it in the Bra-Ket notation?
- ▷ What's the origin and meaning of the Schrödinger equation? How is it related to the non-relativistic energy-momentum relation?
- ▷ What's the difference between classical mechanics and quantum mechanics? Can we derive the laws of classical mechanics using quantum mechanics?
- ▷ What's the origin of the usual quantum operators? (Hint: Read Why $-i\nabla$ is the momentum by Thomas F. Jordan, which is also discussed in Section 3.4. of Ballentine's quantum mechanics book.)

10.4 Map of Quantum Mechanics

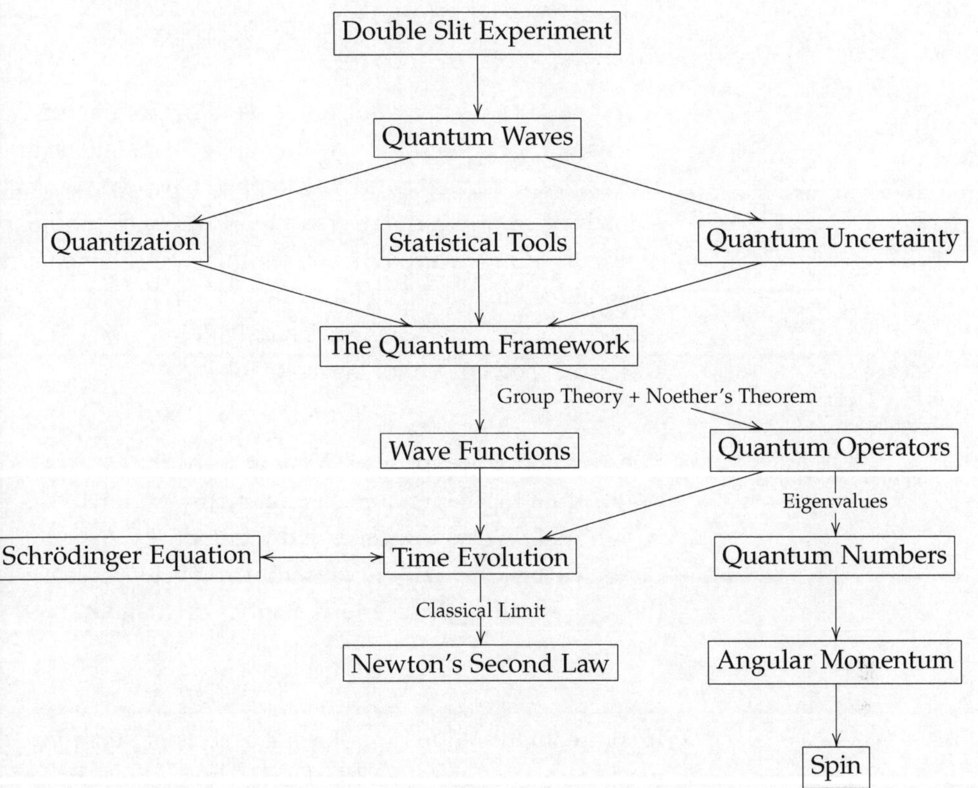

10.5 How To Level Up

If you want to learn more about quantum mechanics, be warned that there is an incredibly broad range of interesting more advanced topics, and no one can master them all. So, if you want to keep focusing on quantum mechanics, try to focus on one advanced topic at a time. Otherwise, it's too easy to get demotivated. Also, for each of these topics, there are certain books and resources that cover them especially well.

Advanced concepts worth your focus:

- ▷ If you want to deepen your understanding of quantum mechanics, you should learn alternative formulations of it. The deepest insights usually happen when you look at something from multiple perspectives. The most popular alternative to the wave function formulation is the path integral formulation. The best way to learn it is to read the book "Quantum Mechanics and Path Integrals" by Richard P. Feynman and Albert R. Hibbs.[6]

[6] But also take note that there is a phase space formulation and a pilot wave formulation of quantum mechanics.

- ▷ The best book to learn how many of the weirdest aspects of quantum mechanics are experimentally explored is "Quantum Theory: Concepts and Methods" by Asher Peres. This is especially important if you want to start thinking about possible interpretations of quantum mechanics.

- ▷ If you're interested in more formal aspects of quantum mechanics, try "Quantum Mechanics" by Leslie E. Ballentine. Other great books that focus on mathematical aspects are "Quantum Theory, Groups and Representations" by Peter Woit, "Quantum Theory for Mathematicians" by Brian Hall, and "An Introduction to the Mathematical Structure of Quantum Mechanics" by F. Strocchi.

- ▷ Of course, it also makes sense to spend some time to learn about the history of quantum mechanics. This is not only a great story but also helps to understand quantum mechanics from various perspectives. A great book to learn more about the exciting history of quantum mechanics is The Quantum Story by Jim Baggott.

▷ An extremely delicate topic is the interpretation of quantum mechanics. Often, students are actively discouraged to spend any time thinking about it because it's easy to spend weeks or even months reading and thinking about it and still get nowhere. But this shouldn't stop you. After all, our goal in physics is not only to describe what is happening but to understand why. And we need more smart people thinking about fundamental issues like the interpretation of quantum mechanics. A good starting point is the book "Interpretation of Quantum Mechanics" by Roland Omnes.

▷ A fun alternative perspective on quantum mechanics that allows you to understand it as a certain generalization of probability theory is presented in the Quantum Computing Since Democritus lectures by Scott Aaronson.[7]

[7] The relevant lecture is available online at `http://www.scottaaronson.com/democritus/lec9.html`.

11

Quantum Field Theory

At A Glance

Why study it?

It's the best theory of elementary particles that we have. Quantum field theory allows us to describe three of the four known fundamental interactions perfectly.

Recommended book:

Student-Friendly Quantum Field Theory by Robert D. Klauber

Main learning objectives:

To understand how we can describe particle scattering processes. By shooting particles onto each other, we can collect facts about nature at the smallest scales and quantum field theory is ideally suited for this task.

11.1 Why Learn Quantum Field Theory?

> Quantum field theory is literally the language in which the laws of Nature are written. **David Tong**

Quantum field theory is the best theory of elementary particles and their interactions that we have. Using a small set of input parameters, quantum field theory allows us to make a large number of wide-ranging predictions. Most importantly, all predictions made using it (so far) have proven to be correct.

Using quantum field theory we can understand the following:

- ▷ The common origin of three of the four known fundamental forces and their properties
- ▷ Where the masses of elementary particles come from
- ▷ Why all electrons we have observed so far "look" exactly the same.

In addition, learning quantum field theory makes sense because it's an extremely flexible tool and can also be used to describe a variety of other things, for example, condensed matter systems or even financial markets.

Before we talk about how you should learn quantum field theory in concrete terms, it makes sense to discuss briefly what it is all about.

11.2 Bird's Eye View of Quantum Field Theory

In quantum field theory particles are excitations of quantum fields and interactions are mediated by a special type of particles called bosons. When we shoot two or more particles onto each other, we can understand what happens by imagining that a boson moves between the locations of the particles. It then causes excitations in possibly different quantum fields which corresponds to the creation of new particles. We can calculate the probability of all possible results of such a collision using perturbation theory that can be visualized using so-called Feynman diagrams.

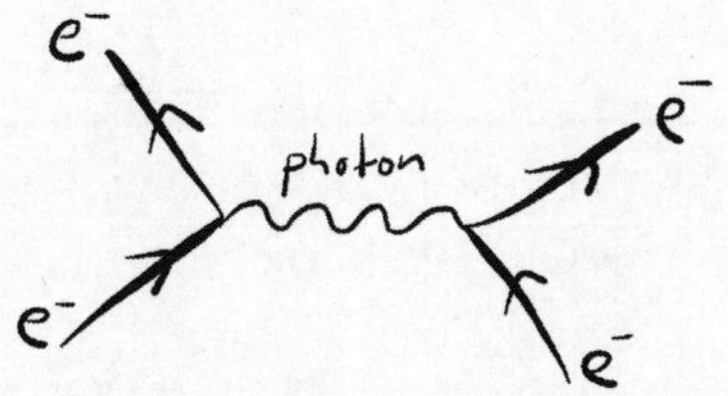

The first magical feature of quantum field theory is how we can use it to describe particles. This is far from trivial since a field and a particle are somewhat contrary concepts. As we have discussed earlier, while a field spreads out everywhere in space, a particle is strictly localized. The key idea in quantum field theory is that particles can be understood as excitations of fields.

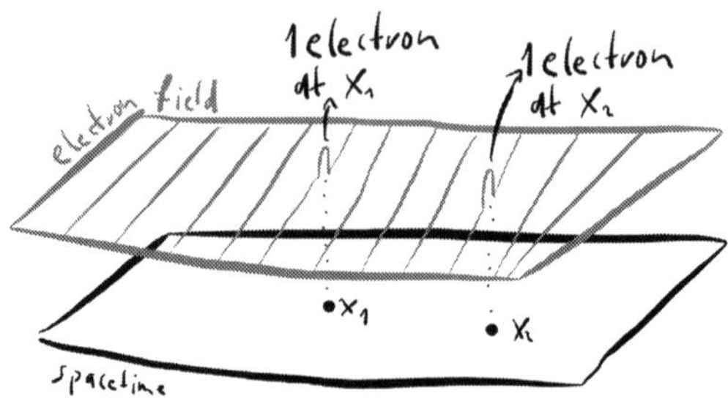

This perspective comes about if we recall that a field is simply a collection of infinitely many harmonic oscillators. At each possible location, we have exactly one harmonic oscillator.

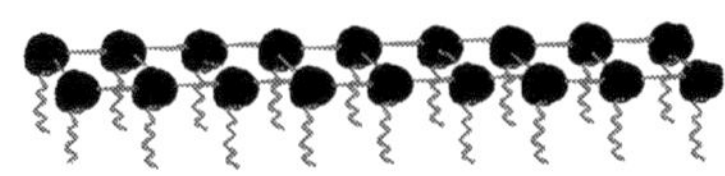

From quantum mechanics, we know that excitations of a harmonic oscillator are quantized, meaning that only a discrete set of excitations is allowed.

Combining these facts, we can conclude that in a quantum field theory we get quantized excitations of our fields. These quantized excitations are what we call particles. For each particle there is an accompanying quantum field. For the electron there is an electron field, for the photon there is the electromagnetic field, and for the Higgs boson there is the Higgs field.

The second crucial idea is that we can reveal deeper facts about nature by smashing particles onto each other. If we accelerate our particles sufficiently before the collision, they "break" and we can get a glimpse of what lies beyond.[1]

[1] Particles don't really break because we understand them as excitations of quantum fields. Therefore, what this really means, is that if the energy of the colliding particles is sufficiently high, we excite additional quantum fields. And these new excitations show up as new particles in our detectors. Since it costs energy to excite quantum fields, it's possible that the original quantum field excitations (the colliding particles) vanishes.

For this reason, a large part of quantum field theory is a

set of tools to describe scattering processes. Here, we start with a given set of particles that we prepare in a collider experiment. For example, two electrons. Then our task is to calculate the probability for a given result after the collision.; for example, two electrons with a different momentum or two different particles like two muons. The main tool that allows us to calculate these probabilities is the time-evolution operator. We evolve our initial state in time and then project out the probability amplitude for our end state in question.

However, while this may sound simple, in practice it isn't. In fact, we can't calculate the probabilities exactly but have to use a perturbative approach. This means, we rewrite our time-evolution operator in terms of a Taylor series and then evaluate the terms individually. Each term in this perturbation series can be visualized using a specific Feynman diagram. Each part of the diagram represents a specific mathematical object in our term. To calculate the contributions from the individual terms in the perturbation series (= the various Feynman diagrams), we must solve the integrals and rewrite the various objects (mainly gamma matrices) appropriately.

11.3 How To Understand The Fundamentals

By far the best quantum field theory textbook is "Student-Friendly Quantum Field Theory" by Robert D. Klauber[2].

[2] Robert Klauber. *Student friendly quantum field theory : basic principles and quantum electrodynamics.* Sandtrove Press, Fairfield, Iowa, 2014. ISBN 9780984513956

Don't let the somewhat weird cover fool you. It's a serious, non-crackpot textbook that covers all relevant equations. But it presents them as gently as possible and always makes sure the reader doesn't get lost. For example, there are lots of overview charts that help to understand how each concept and equation fits into the bigger picture.

Most importantly, Klauber always explains where each concept comes from and why it makes sense to use it in the context at hand.

Additional good introductory textbooks are "Introduction to Elementary Particles by Griffith"[3], Quantum Field Theory and the Standard Model by Schwartz[4] and "An Invitation to Quantum Field Theory" by Álvarez-Gaumé and Miguel Vázquez-Mozo[5].

[3] David Griffiths. *Introduction to elementary particles.* Wiley-VCH, Weinheim Germany, 2008. ISBN 9783527406012

[4] Matthew Schwartz. *Quantum field theory and the standard model.* Cambridge University Press, Cambridge, United Kingdom New York, 2014. ISBN 9781107034730

[5] Luis Alvarez-Gaumé and Miguel Vázquez-Mozo. *An invitation to quantum field theory.* Springer, Heidelberg New York, 2012. ISBN 978-3-642-23728-7

The standard references that professors love to recommend are An Introduction to Quantum Field Theory by Peskin and Schröder and The Quantum Theory of Fields Vol. 1 by Weinberg. But you should consult both only sporadically, especially as a beginner, because they oftentimes make quantum field theory appear a lot more complicated than it really is. Both make it too easy for a beginner to get lost in the details and it's really hard to keep track of which derivations are essential.

Key concepts to focus on:

- ▷ The basic equations, their solutions and interpretation: the Klein-Gordon equation, Dirac's equation, Maxwell's equations and the Proca equations
- ▷ The canonical commutation relations and their connection to how our particle interpretation comes about
- ▷ Creation and annihilation operators
- ▷ The time evolution operator
- ▷ The meaning and role of propagators
- ▷ Spinors and the gamma matrix algebra
- ▷ Polarization vectors
- ▷ The perturbation series, Feynman diagrams and Feynman rules

Example questions:

- ▷ What do Feynman diagrams really represent?
- ▷ What's the common origin of the Klein-Gordon equation, Dirac's equation and the Maxwell equations?
- ▷ In what sense is a field a set of an infinite number of harmonic oscillators and how does this lead to the emergence of particles in a field theory?
- ▷ You should try to understand the following two statements: "One could say that a spinor is the most basic sort of mathematical object that can be Lorentz-transformed"[6] and "a spinor is the square root of a vector".

[6] Andrew M. Steane. An introduction to spinors, 2013

- ▷ In what sense does (gauge) "symmetry dictate interactions"? In what sense is quantum field theory the unique fundamental theory we end up with under reasonable assumptions? (Hint: This is discussed extensively in Weinberg's Quantum Field Theory books.)

▷ Why do we get great results by calculating the first few terms in the perturbation series even though if we sum up all terms we get infinity?

▷ What kinds of infinities do show up in quantum field theory and why?

▷ Why is there no macroscopic electron field while there is a macroscopic photon field (=the electromagnetic field)? (Hint: Pauli exclusion principle.)

▷ In what sense can we say that quantum mechanics is simply quantum field theory in 0+1 dimensions?

▷ If gauge symmetry is not really a symmetry, how can it play such an important role in the Higgs mechanism?

▷ In what sense is statistical mechanics a Wick rotated version of quantum field theory?

11.4 Map of Quantum Field Theory

Formalism

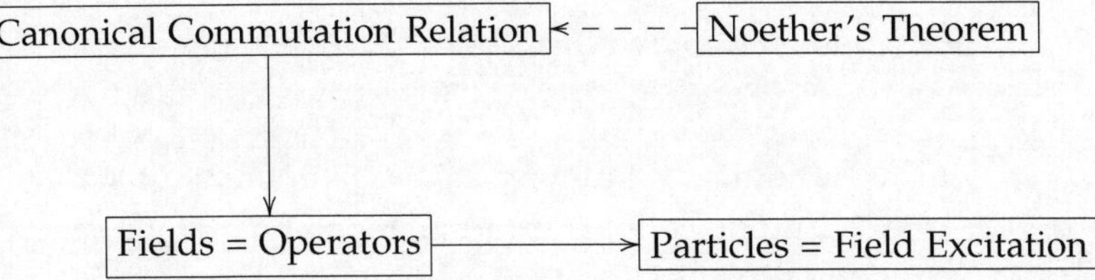

Ingredients

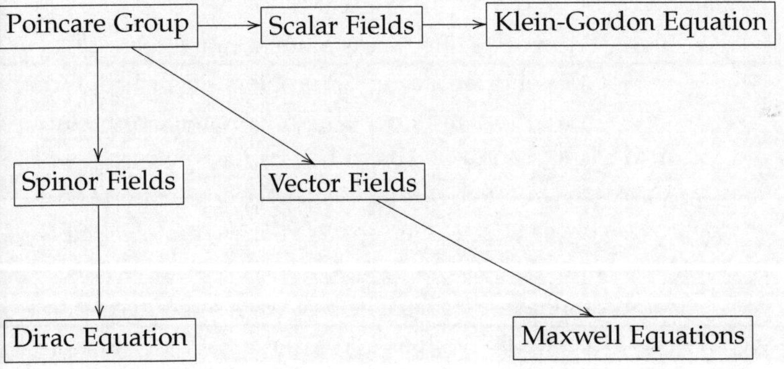

11.5 How To Level Up

Once you've mastered the fundamentals, you can decide whether you want to dive deeper or if you want to move on to another subject like general relativity (as one example).

Advanced concepts to focus on:

- ▷ The path-integral formulation of quantum field theory: the second most popular formulation of quantum field theory right after the standard canonical formulation. Like any theory in physics, there isn't one unique way to use it in practice. It's useful to study, for example, global properties of quantum systems instead of only local ones. To learn it, try "Quantum Field Theory in a Nutshell" by Zee and "Quantum Field Theory" by Srednicki.
- ▷ Non-perturbative effects: instantons, sphalerons, monopoles and the strong CP problem. It turns out that lots of important and interesting phenomena in quantum field theory can not be described using a perturbative approach. In other words, the usual perturbation series misses these effects. One spectacular example is that we now know that thanks to such non-perturbative effects the proton is not stable. The best book to go deeper in this direction is "Solitons and Instantons" by R. Rajaraman. Also try "Topological Solitons" by Manton and Sutcliff, "Classical Solutions in Quantum Field Theory" by E. J. Weinberg, "Quarks, Leptons and Gauge Fields" by K. Huang and "Advanced Topics in Quantum Field Theory" by M. Shifman.
- ▷ The renormalization group: is one of the greatest inventions ever made in physics. In some sense, without it we wouldn't be able to understand why physics works at all, i.e. why we can describe what happens using relatively

simple rules even though we don't know all microscopic details. Often, the renormalization group is discussed solely in the context of quantum field theory, but it really has an incredibly broad range of applications. The renormalization group is a general tool that allows us to zoom out from a given microscopic description and derive an oftentimes simpler description that is valid if we look at the system from a distance. The standard story in the context of quantum field theory is covered in almost every textbook, like in Chapter 8 of "An Invitation to Quantum Field Theory" by Álvarez-Gaumé and Miguel Vázquez-Mozo. But the best way to understand it in a broader context is the free online course titled "Introduction to Renormalization" by Simon DeDeo.[7] A great book on the topic is Elements of "Phase Transitions and Critical Phenomena" by Nigel Goldenfeld

[7] https://www.santafe.edu/engage/learn/courses/intro_renorm

▷ Differential geometry of gauge theory: this allows us to understand all modern theories of physics from a common perspective. Great books are "Topology, Feometry, and Gauge Fields" by Naber and "Gauge Fields, Knots and Gravity" by Baez and Muniain.

▷ Anomalies: a rough definition of an anomaly is that we expect that our model has a certain symmetry by looking at the Lagrangian, but then we notice that this symmetry does not survive the quantization procedure. There are various types of anomalies and some of them are useful to check the consistency of a given model, while others allow for more speculative mechanisms like Leptogenesis, as one example. A gentle introduction can be found in Chapter 9 of "An Invitation to Quantum Field Theory" by Álvarez-Gaumé and Miguel Vázquez-Mozo.

▷ Calculations beyond tree-level. This is an infinitely deep rabbit hole and there is a worldwide community of professional physicists who focus solely on the development of new methods to calculate higher orders in perturbation theory ("multi-loop calculations"). If you want to get a glance of what this business is all about, have a look at "Analytic Tools for Feynman Integrals" by Smirnov.

▷ Quantum field theory in mathematically (more) rigorous terms. There is still an open $1,000,000 Millennium Problem related to the rigorous construction of gauge models in quantum field theory. So, there remains a lot to be done. But, of course, quantum field theory doesn't have to be as sloppy as physicists usually make it to be and lots of progress towards a more rigorous formulation has been made in the past decades. A gentle first step in this direction is "Quantum Field Theory: A Modern Perspective" by Nair.

▷ Spontaneous symmetry breaking and the Higgs mechanism. These are essential ideas at the heart of the best model of elementary particles that we have (the Standard Model). You can find discussions in any modern textbook. Try, for example, Chapter 11 in "A Modern Introduction to Quantum Field Theory" by Maggiore.

▷ The CPT theorem and the spin-statistics theorem. These are really hard topics but any competent physicists should know the main message of these theorems and their underlying assumptions. Good discussions can be found in "Quantum Field Theory and the Standard Model" by Schwartz.

▷ The loop formulation of gauge field theory, which has nothing to do with the multi-loop calculation mentioned above. Instead, the main idea is to replace the usual notions we use to describe quantum field theory with a new set of notions that make it easier to understand the underlying physics. A really nice and short book on this perspective on gauge theory is "Some Elementary Gauge Theory Concepts" by Tsou and Chan.

12

General Relativity

At A Glance

Why study it?
It's the best model of gravity that we have.

Recommended book:
Relativity, Gravitation and Cosmology by Ta-Pei Cheng

Main learning objectives:
Grasp where Einstein's equation comes from and in what sense gravity can be understood as a feature of spacetime.

12.1 Why Learn General Relativity?

"General relativity is the most beautiful physical theory ever invented." **Sean M. Caroll**[1]

[1] Sean Carroll. *Spacetime and geometry : an introduction to general relativity*. Pearson, Harlow, Essex, 2014. ISBN 9781292026633

Gravity is one of only four known fundamental interactions. And Einstein's general relativity is the best model describing gravity that we have. So far, all experimental predictions made using have it turned out to be correct. A famous example is the recent discovery of gravitational waves.

But general relativity is also interesting because of its unique origin story. In some sense, Einstein's model of gravity is a true unicorn in physics. It was invented purely through deep thinking. In contrast, all other models and theories that we have were slowly built up using lots of experimental puzzle pieces. Einstein bypassed this oftentimes slow feedback loop and jumped directly to the right conclusions. For this reason, general relativity is extremely popular among theoretical physicists and is for many the prototypical example how physics can advance in the absence of experimental progress. In addition, general relativity is regarded by many as the most beautiful physical model ever invented.

Another reason to spend time thinking about Einstein's ideas is that there are still lots of unresolved mysteries directly connected to gravity. The most famous problem is that there is no generally accepted quantum model of gravity ("quantum gravity"). A fundamental requirement for any such model is that it must yield Einstein's model on large scales. Therefore, a solid understanding of general relativity is essential if you want to think about quantum gravity. Similarly, Einstein was only able to derive the correct model because he knew that it had to yield Newton's model of gravity for all systems in which gravity is weak.

Now, before we talk about how you should learn general relativity in concrete terms, it makes sense to discuss briefly what it is all about.

12.2 Bird's Eye View of General Relativity

The basic idea at the heart of general relativity is that massive objects curve spacetime and all objects that move through spacetime always follow the "shortest paths". But since spacetime is now curved, these paths are no longer straight lines. As a result, we can observe that objects get deflected if massive objects are present and this deflection is what we usually attribute to the force called gravity. John Wheeler famously put it this way:

> "Spacetime tells matter how to move; matter tells spacetime how to curve. "

The interplay between energy, momentum and spacetime is described by the Einstein equation:

$$\underbrace{T_{\mu\nu}}_{\text{energy-momentum tensor}} = C \underbrace{G_{\mu\nu}}_{\text{Einstein Tensor}} . \tag{12.1}$$

The constant C encodes how strongly spacetime reacts to the presence of energy-momentum.[2]

[2] Specifically, $C = \frac{8\pi G}{c^4}$ where G is Newton's gravitational constant and c the speed of light.

On the right-hand side of Einstein's equation, we have a purely geometric object that encodes properties of spacetime. On the left-hand side, we have the energy-momentum tensor that encodes the presence of objects. In words, a basic statement encoded in Einstein's equation is that whenever there is a nonzero energy-momentum density in the system, spacetime will be curved. Or to put it another way, the main feature of general relativity is that spacetime is no longer a boring flat background thing, but is directly influenced by all objects. In turn, these objects are influenced by the geometry of spacetime.

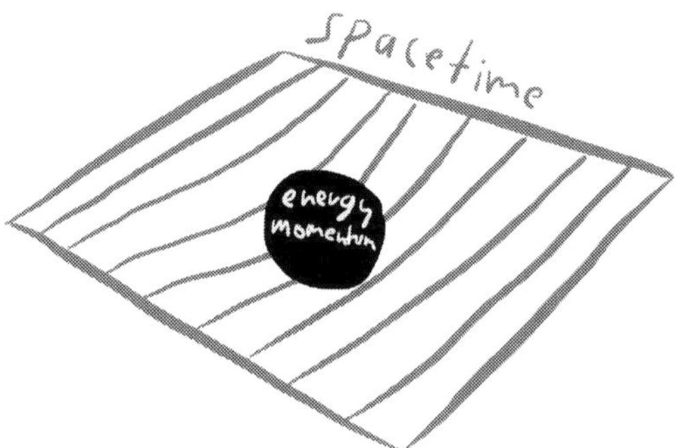

The main ingredient that we use to describe properties of spacetime is called the metric. A metric encodes the distance between two given points. Using the metric, we can construct more advanced quantities like the curvature tensor for example. The curvature tensor tells us whether or not spacetime is curved. Einstein's equation allows us to derive how exactly spacetime is curved if we provide how energy and momentum are distributed throughout the system. However, in practice this is an extremely difficult task and can only be carried out analytically for highly idealized situations.

Finally, we can calculate how test masses move through any given curved spacetime configuration by using the geodesics equation. Geodesics are the "shortest" paths between two points in a flat spacetime configuration and the straightest path between two points in a curved spacetime configuration. For example, on a sphere the geodesics are "great circles".

12.3 How To Understand The Fundamentals

My favorite general relativity textbook is "Relativity, Gravitation and Cosmology" by Ta-Pei Cheng[3]. In contrast to almost all other books on general relativity, it's a quick read. But still it explains all fundamental aspects carefully and clearly. This makes Cheng's book an ideal starting point for further studies.

[3] T.P. Cheng. *Relativity, gravitation and cosmology : a basic introduction*. Oxford University Press, Oxford New York, 2010. ISBN 9780199573646

But before reading that book, it makes sense to read a short primer titled "A No-Nonsense Introduction to General Relativity" by Sean M. Caroll.[4]

[4] `https://preposterousuniverse.com/wp-content/uploads/2015/08/grtinypdf.pdf`

Additional good introductory textbooks are "Gravity" by James B. Hartle[5], "Einstein Gravity in a Nutshell" by A. Zee[6] and "A First Course in General Relativity" by Bernard Schutz[7].

[5] [Hartle, 2003]

[6] [Zee, 2013]

[7] [Schutz, 2009]

Key concepts to focus on:

- ▷ Einstein's equation
- ▷ The geodesics equation
- ▷ The metric, curvature, connection and covariant derivatives.

Example questions:

- ▷ Is spacetime a dynamic object?
- ▷ Which features of Einstein's equation make it so hard to solve?
- ▷ In what sense is Newton's model of gravity included in Einstein's model?
- ▷ Which features of general relativity are completely analogous, or at least very similar, to gauge models in quantum field theory?
- ▷ Why is it commonly believed that the particle mediating gravitational interactions is a spin-two particle (the graviton), while all other fundamental interactions are mediated by spin 1 particles? (Hint: A great explanation is provided in Zee's "Quantum Field Theory in a Nutshell" book.)
- ▷ In what sense does "general relativity behave like the square of Yang-Mills gauge theory"?
- ▷ A manifold cannot only possess a nonzero curvature but also a nonzero torsion and a nonzero non-metricity. Is it also possible to construct theories where torsion or non-metricity are dynamical objects? (Hint: for a nice exploration of such ideas see "Nonmetricity Formulation of General Relativity and its Scalar-Tensor Extension" by Järv et. al. and "Gauge Theory of Gravity and Spacetime" by Friedrich W. Hehl.)

12.4 Map of General Relativity

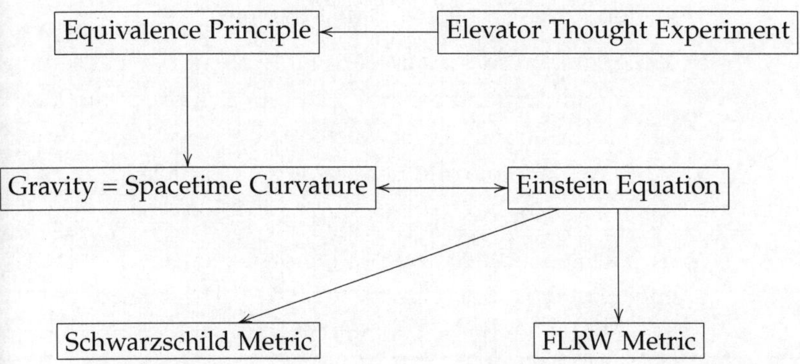

12.5 How To Level Up

Once you've mastered the fundamentals you can decide which aspects of general relativity inerest you the most and then dive deeper in.

But first of all, to deepen your understanding of general relativity you should read The Meaning of Einstein's Equation by John C. Baez and Emory F. Bunn.

The best general books to read deep explanations of advanced aspects of general relativity are Gravitation by Charles W. Misner, Kip S. Thorne and John Archibald Wheeler, General Relativity by Robert M. Wald and Gravitation and Cosmology by Steven Weinberg.

Advanced concepts worth focussing on

▷ A great book to learn how we can describe general relativity in more abstract terms and how this helps to understand the connection to all other models of fundamental interactions is Gauge Fields, Knots, and Gravity by John Baez and Javier P. de Muniain.

▷ Other great resources to learn more about in what sense we can understand general relativity as a gauge model are Gauge Gravity: a forward-looking introduction by Andrew Randono and Gravity from a Particle Physicists' perspective by Robert Percacci.

▷ To learn about the various attempts to construct a viable model of Quantum Gravity, start with Three Roads to Quantum Gravity by Lee Smolin. It provides a perfect overview and is quick read. Afterwards, try Quantum Gravity by Rovelli which focusses on the Loop Quantum Gravity perspective and Quantum Gravity by Claus Kiefer which discusses many general aspects and String Theory.

▷ Another important aspect is in what sense we can understand general relativity as a quantum effective field theory. A good starting point is Introduction to the Effective Field Theory Description of Gravity by John F. Donoghue.

▷ The usual interpretation of general relativity is that gravity is simply the result of the curvature of spacetime and in this sense there is no gravitational field which lives on top of this structure. But it is also possible to turn this interpretation completely upside down. We can say that there is no spacetime, only the gravitational field. In other words, spacetime is merely an illusion which results from the gravitational interactions which happen all the time. For an extremely lucid discussion of this point of view, see Section 2.2.5 in Rovelli's Quantum Gravity book.[8]

[8] I summarized some of the main ideas at http://jakobschwichtenberg.com/the-true-magic-hidden-inside-general-relativity/.

13

Statistical Mechanics and Thermodynamics

I'm not an expert in neither thermodynamics nor statistical mechanics and I don't play one on the internet or otherwise.

I simply don't know enough about these topics to give any substantial recommendations. While it would be easy to add a few recommendations and smart sounding sentences based on second-hand information, this would be completely against the spirit of this book. I hope that I will develop a better understanding of thermodynamics and statistical mechanics at some point in the future. As soon as this is the case, I will modify this chapter accordingly.

But let me assure you that my ignorance should be by no means an indicator that thermodynamics and statistical mechanics are somehow less important. Just listen to Albert Einstein who once remarked that thermodynamics

> "is the only physical theory of universal content, which I am convinced, that within the framework of applicability of its basic concepts will never be overthrown."

Moreover, John Goold et. al. came up with an analogy that nicely illustrates the unique status of thermodynamics[1]:

[1] John Goold, Marcus Huber, Arnau Riera, Lídia del Rio, and Paul Skrzypczyk. The role of quantum information in thermodynamics—a topical review. *Journal of Physics A: Mathematical and Theoretical*, 49 (14):143001, feb 2016. DOI: 10.1088/1751-8113/49/14/143001

> "If physical theories were people, thermodynamics would be the village witch. Over the course of three centuries, she smiled quietly as other theories rose and withered, surviving major revolutions in physics, like the advent of general relativity and quantum mechanics. The other theories find her somewhat odd, somehow different in nature from the rest, yet everyone comes to her for advice, and no-one dares to contradict her."

Moreover, statistical mechanics is the fundamental theory that underlies thermodynamics. It allows us to understand why the almost magical rules of thermodynamics work the way they do.

Hopefully, these comments and quotes sparked your interest and will motivate you to learn more about these exciting topics. To give you at least some orientation, here are a few pointers:[2]

[2] As mentioned above, I'm not an expert and you should take these recommendations with a grain of salt.

▷ The thermodynamics lecture notes by Sanjoy Mahajan[3] contain many nice examples that help to build some intuitive understanding of many of the most important notions and also useful further reading recommendations.

[3] `http://www.inference.org.uk/sanjoy/teaching/thermo/notes.html`

▷ Arguably the most important notion in thermodynamics is entropy. What helped me to understand it properly were:

- The essay on Entropy and Gambling by Brian Skinner.[4]
- Entropy? Honest! by Tommaso Toffoli [5].
- Edwin T. Jaynes' papers The Second Law as Physical Fact and as Human Inference [6] and Gibbs vs Boltzmann Entropies [7].

▷ Thermodynamics and classical mechanics are a lot more similar than one might think at first glance. This is discussed nicely in the blog post titled Classical Mechanics versus Thermodynamics by John Baez.[8]

▷ What is Statistical Mechanics? by Roman Frigg[9] is an excellent starting point to get an overview quickly.

▷ I immensely enjoyed reading the textbook titled Statistical Mechanics by James P. Sethna. Another good textbook is Fundamentals of Statistical and Thermal Physics by Frederick Reif.

Once you've developed some solid understanding, you might enjoy reading about the following topics:

▷ There is an intriguing connection between quantum theories and statistical mechanics. A great starting point is the series of blog posts on "quantropy" by John Baez.[10]

▷ There are interesting attempts to understand the laws of general relativity from a thermodynamical perspective. This idea goes by the name "entropic gravity". A gentle summary of these ideas can be found in the Quanta magazine article "The Case Against Dark Matter" by Natalie Wolchover.[11]

[4] https://gravityandlevity.wordpress.com/2009/04/01/entropy-and-gambling/

[5] Tommaso Toffoli. Entropy? honest! *Entropy*, 18(7), 2016. ISSN 1099-4300. DOI: 10.3390/e18070247. URL https://www.mdpi.com/1099-4300/18/7/247

[6] E. T. Jaynes. The second law as physical fact and as human inference, 1998

[7] E. T. Jaynes. Gibbs vs Boltzmann Entropies. *American Journal of Physics*, 33: 391–398, May 1965. DOI: 10.1119/1.1971557

[8] https://johncarlosbaez.wordpress.com/2012/01/19/classical-mechanics-versus-thermodynamics-part-1/

[9] http://philsci-archive.pitt.edu/9133/

[10] https://johncarlosbaez.wordpress.com/2011/12/22/quantropy/

[11] https://www.quantamagazine.org/erik-verlindes-gravity-minus-dark-matter-20161129/.

Part IV
Further and Meta Subjects

"The best performers observe themselves closely. They are in effect able to step outside themselves, monitor what is happening in their own minds, and ask how it's going."

Geoff Colvin in Talent is Overrated

It's simply not enough to know how to navigate yourself in the landscape of physical theories. There are several other skills and topics that every competent physicist should have some understanding of:

- ▷ The one topic that provides the highest return on investment is **learning how to learn**. Even if you think of yourself as a fairly competent learner, there is always room for improvement and each new idea has the potential to accelerate your progress dramatically.
- ▷ Secondly, you should spend some time to learn **how to think**. There are many incredibly useful methods and learning them early on will give you a real head start.
- ▷ **Mathematics** is the language of the continent of physics. So we need to talk about how you should learn it.
- ▷ **History** can be confusing and the historical path to a theory is usually full of dead ends and misunderstandings. Nevertheless, once you've got a solid understanding of a theory from a modern perspective, it can be extremely useful to learn more about the history of the subject. Understanding how a theory was developed is not only inspiring but often contains many important lessons.
- ▷ It's easy to lose sight of the **big picture**. So from time to time, you should make a conscious effort to understand how the topic you're trying to understand right now is connected to the rest of physics.
- ▷ Physics is still done by humans not robots. So there are always aspects of physics that are not entirely rational and driven by sociological forces. Even as a self-learner, you should try to immunize yourself as much as possible by studying **common fallacies** and **sociological aspects of physics**.
- ▷ Another incredibly important topic that physicist rarely talk about is the **philosophy of science**. Quite shockingly,

philosopher has even become a derogatory term among certain physicists. But if done right, philosophy can be a lot more than fluffy talk. Especially in times in which no one really knows in which direction we should go next, philosophy can offer invaluable guidance.

If you're not a university student, you will probably be surprised to hear that most of these topics are not part of a normal university curriculum. They are rarely ever taught explicitly and students are expected to learn them through osmosis and trial-and-error. This again demonstrates that it's not sufficient to rely on some predefined university curriculum.

Therefore, let's discuss how you can study these topics independently.

14

Learn How to Learn

In Part V, I will share some advice on how to learn physics that I wished I had received at the beginning of my learning journey. There are, however, many other people who have written much smarter things on the general topic of learning how to learn.

In particular, I think that the book Ultralearning by Scott H. Young [1] is brilliant. I read it only after I'd already finished most chapters of Part V. Otherwise, I would have maybe simply referred you to this book because he articulates many ideas and concepts much better than I ever could. He outlines perfectly all the steps and principles that are necessary to self-learn any topic effectively. Moreover, he describes many inspiring examples. In particular, by applying the method he describes he was able to learn the curriculum of MIT's four-year computer science undergrad in a single year.

[1] Scott Young. *Ultralearning: seven strategies for mastering hard skills and getting ahead*. Harper Collins Publ., UK, 2019. ISBN 9780008305703

One key to this project was that he ignored most of the material provided by the university and focused on learning the topics using whatever material suited his needs best. In other words, instead of following the curriculum religiously, he carved out his own path and this allowed him to succeed in a much shorter amount of time. This idea will also be one of the main themes in Part V. A second key is to stop waiting till you feel ready. Instead, you need to immerse yourself in the topic, jump right in and learn the topic aggressively. The combination of aggressiveness and self-directed learning is what Scott H. Young calls ultralearning.

To show what is possible with this approach he recently taught himself quantum mechanics in a month-long project and at the end of it, successfully passed the final exam for MIT's first quantum mechanics class, 8.04 – Quantum Physics I.[2]

[2] https://www.scotthyoung.com/blog/2019/05/02/qm-project-complete/

A second extremely useful resource is the free Coursera class titled "Learning How to Learn: Powerful mental tools to help you master tough subjects".[3]

[3] https://coursera.org/learn/learning-how-to-learn

Spending a few days studying these resources (plus Part V) will give you all the tools that you need to learn physics effectively.

15

Learn How to Think

It's tempting to believe that brilliant thinkers are born that way. This, however, couldn't be further from the truth. Of course, there are genetic outliers. But most "brilliant" people have simply learned to use their mind differently. They've learned thinking strategies and developed habits that allow them to see things differently.

The Indian philosopher Jiddu Krishnamurti summarized it nicely:

> "The trouble with our system of upbringing is that it teaches us what to think not how to think."

Students are not only expected to pick up learning strategies by trial-and-error, but are often also never taught effective thinking strategies. For some reason, professors all around the world expect their students to learn this through osmosis. Even worse, most students aren't aware that this is something you can learn and improve.

This is crazy. Learning how to think is arguably the second best investment of your time right after learning how to learn. In addition, smart thinking is probably the skill that physicists are most famous for.

If you only spend a few days studying the skill of thinking, you will be able to solve problems in minutes that otherwise take hours. Moreover, you will make fewer mistakes and start to enjoy hard challenges. A regular student reaches this level only after years of osmosis learning and some never do.

There are, of course, many aspects of a skill like thinking. But the one aspect with the quickest return on investment is a solid understanding of general hands-on problem-solving skills.

A great book that carefully explains and motivates all the problem-solving skills that you need to get started in physics is:

- "Street-Fighting Mathematics: The Art of Educated Guessing and Opportunistic Problem Solving" by Sanjoy Mahajan.

Sanjoy Mahajan is a master in making tacit knowledge explicit. So if you want to dive deeper, you should also give his book The Art of Insight in Science and Engineering a try.

In addition to directly applicable problem-solving methods, there are also more abstract aspects of your thinking that you can improve which, in the long term, potentially yield even greater payoffs. This is what we will talk about in the following two sections.

15.1 Biases and Fallacies

Physics is (still) a human endeavor and humans often think and act irrationally.[1] The best way to describe all of this is by using the terms logical fallacy and cognitive bias. A cognitive bias is something that exists in a mind, while a logical fallacy is something that exist in an argument. So when you read a faulty argument, you may fall for its fallacies because a particular cognitive bias clouds your vision.

[1] Two great books on this topic are

Dan Ariely. *Predictably irrational : the hidden forces that shape our decisions*. Harper Perennial, New York, 2010. ISBN 9780061353246; and Daniel Kahneman. *Thinking, fast and slow*. Farrar, Straus and Giroux, New York, 2011. ISBN 9780141033570

Unfortunately, many physicists don't acknowledge the existence of these thinking errors and act as if physicists were too smart to fall victim to logical fallacies and cognitive biases. Intelligence simply doesn't protect you from thinking biases.

A rare exception was, once again, Richard Feynman who freely admitted:

> "The first principle is that you must not fool yourself and you are the easiest person to fool."

Be assured that everyone, no matter how smart, is regularly influenced by things that aren't entirely rational. This is simply human nature and a result of how we evolved evolutionary.

Fallacies that are quite common in physics are:

▷ Arguments from popularity. This fallacy happens whenever someone declares that something is true because the great majority of people in general agree with some particular position. The popularity of an idea, however, has absolutely no bearing on its validity. For example, the majority of people believed for centuries that the earth is flat. Or to give another less trivial example, nowadays most physicists believe that the fundamental constants are, well, constant in time. This may very well be true but it's simply not valid to declare that a theory with varying fundamental constants is wrong because the majority of physicist believe that fundamental quantities like the speed of light should be constant in time.[2]

▷ Arguments from authority. Just as it's not a valid argument to declare that something is true because everyone claims that it is, it's not sufficient to cite some authority.[3] Experts can be wrong just as everyone else.[4]

For example, 1983 Stephen Hawking gave a lecture on cosmology, in which he explained[5]

> "However, if there are two facts about our universe which we are reasonably certain, one is that it is not exponentially expanding and the other is that it contains matter."

Only 15 years later, physicists were no longer "reasonably certain" that the universe isn't exponentially expanding.

On the contrary, we are now reasonably certain of the exact opposite. By observing the most distant supernovae two experimental groups established the accelerating expansion as an experimental fact. This was a big surprise for everyone and rightfully led to a Nobel Prize for its discoverers.

[2] An entertaining description of why it may be interesting to think about fundamental constants that vary in time and, more importantly, the sociological implications of such a proposal can be found in the book Faster Than The Speed Of Light by Joao Magueijo [Magueijo, 2004]. (This is by no means an endorsement of Magueijo's theory. But it's certainly an interesting case study.)

[3] You've probably noticed by now that I'm guilty of quoting famous people. I do this primarily, however, because often a famous person formulated an idea much better than I ever could. You shouldn't believe in these statements because of the person's fame or authority. Ultimately, as always, you need to decide for yourself what is really true and what isn't..

[4] This is particularly true if experts comment on things outside of their original domain of expertise. For example, there is no reason to believe that a Nobel price winner in physics is someone whose nutrition advice you should believe.

[5] Lizhi Fang and Remo Ruffini. *Quantum cosmology.* World Scientific, Singapore Teaneck, NJ, USA, 1987. ISBN 978-9971502935

The Russian mathematician Vladimir Voevodsky, after he had noticed that many ground-breaking papers contained errors, famously summarized:[6]

> "A technical argument by a trusted author, which is hard to check and looks similar to arguments known to be correct, is hardly ever checked in detail."

This is true not just in mathematics, but also in physics. Physicists regularly cite authorities even though they haven't checked all the details themselves.

[6] `https://www.ias.edu/ideas/2014/voevodsky-origins`

So in summary, to be sure of something you need to verify for yourself whether or not there is a logical chain of arguments that leads to a particular result. You must learn to recognize when you jump too quickly to a certain conclusion. And you must also be able to recognize when others do this.

Only when you understand logical fallacies, you can "start to see things how they really are, not how you think they are"[7].

[7] Bo Bennett. *Logically fallacious : the ultimate collection of over 300 logical fallacies*. Archieboy Holdings, LLC, Sudbury, MA, 2015. ISBN 978-1456607524

Maybe it's even more important to understand why people regularly fall for such fallacies. In my experience, the following cognitive biases play a critical role in physics:

▷ Communal reinforcement (also known as group think or bandwaggoning). This bias happens whenever the members of a group reassure each over constantly that what they are doing is the right thing. It works because, unfortunately, repetition is a convincing argument. Moreover, members of such a group will often use arguments from popularity.

▷ Irrational escalation. This bias happens whenever someone cannot give up because he has already invested a lot. This could be money for a gambler or time spent investigating a particular theory for a physicist.

▷ Wishful thinking and confirmation bias. We are a lot more likely to believe that something is true if we want it to be true. Moreover, if we want something to be true, we tend to focus on information that confirms our beliefs. Upton Sinclair famously observed:

> "It is difficult to get a man to understand something, when his salary depends upon his not understanding it."

There are, of course, many additional fallacies and biases you should be aware of and you can find many comprehensive lists online.[8] To start with, you should spend at least an hour or so familiarizing yourself with the most important ones.

[8] You can find a nice overview at `http://www.csun.edu/~dgw61315/fallacies.html` and at `http://www.fallacyfiles.org/howtouff.html`. An important additional fallacy that is rarely ever discussed is the mind projection fallacy. This fallacy was popularized by Edwin T. Jaynes who summarized it as follows: "Failure to see the distinction between reality and our knowledge of reality puts us on the Royal Road to Confusion."

Be warned, however, that knowing about logical fallacies and biases does not make you immune to them. But at least, it's a start.

If you're willing to invest more time and really want to step up your thinking game, you should read

▷ Seeking Wisdom by Peter Bevelin [9] and

▷ Pragmatic Thinking and Learning by Andy Hunt [10].

[9] Peter Bevelin. *Seeking wisdom : from Darwin to Munger*. PCA Publications, Malmoe, Sweden, 2013. ISBN 978-1578644285

[10] Andrew Hunt. *Pragmatic thinking and learning : refactor your "wetware*. Pragmatic, Raleigh, 2008. ISBN 978-1934356050

Moreover, if you want to understand which role social aspects play in physics, you should study, well, the sociology of physics. We will talk about this in Chapter 19. In addition, being able to think clearly about a scientific topic requires at least *some* meta understanding of what you're trying to achieve and which general methods exist to verify your ideas. Speaking broadly, these topics fall under the umbrella called "the philosophy of science", which is discussed in Chapter 18.

In the following chapter, we will talk about mathematics. At this point, you are probably wondering why mathematics hasn't been discussed in any detail although it's undeniably the language of physics. Hopefully, this will become clear in the next few minutes.

16

Mathematics

For most beginners it doesn't make sense to learn mathematical tools in advance. Without knowing the physical context it is extremely hard to judge which concepts are really important. Mathematics is, literally, an infinite rabbit hole. There are infinitely many fundamental problems and no objective way to decide what is fundamental and what isn't. Anyone can come up with some set of axioms and then spend a lifetime studying their implications. In addition, once you'll get to a point where you need some specific mathematical concept for a physics problem, you'll have forgotten all the details anyway. So in other words, just-in-case learning is not a good approach when it comes to mathematics because it's far too easy for beginners to get lost. It's smarter to focus on mathematical concepts whenever you need them concretely for a physics problem. The main advantage of this just-in-time learning approach is that you'll always know the proper motivation for each mathematical definition and it's far easier to understand mathematical ideas if you have a concrete picture in your mind.

For example, you'll learn:

▷ Calculus and how to solve differential equations in the context of Newtonian mechanics.

▷ Variational calculus in the context of Lagrangian mechanics.

▷ Vector and tensor calculus and, if you're feeling adventurous, differential geometry in the context of electrodynamics.

▷ Group theory in the context of special relativity.

▷ Complex analysis in the context of quantum mechanics.

Almost all books recommended in the subject guides explain the essential mathematical tools sufficiently well. Of course, once you've mastered the fundamentals it can make sense to dive deeper and learn more about the mathematical machinery by picking up a dedicated math book.

Be warned, however, that most math books are not really suited for physicist. They mostly contain definitions, theorems and proofs with rarely any explanation why you should care about the concepts and how they can be used to describe things in the real world.[1] Vladimir Arnold once summarized the current state of the mathematical literature perfectly:[2]

> "It is almost impossible for me to read contemporary mathematicians who, instead of saying 'Petya washed his hands,' write simply: 'There is a $t_1 < 0$ such that the image of t_1 under the natural mapping $t_1 \to \text{Petya}(t_1)$ belongs the set of dirty hands, and a $t_2, t_1 < t_2 \leq 0$ such that the image of t_2 under the above-mentioned mapping belongs to the complement of the set defined in the preceding sentence."

[1] The following mantras, which is often attributed to John Von Neumann, is quite popular: "In mathematics you don't understand things. You just get used to them."

[2] The Mathematical Intelligencer, December 1987, Volume 9, Issue 4, pp 28–32.

Or as Nobel laureate Chen Ning Yang once observed

> "There are only two kinds of math books: Those you cannot read beyond the first sentence, and those you cannot read beyond the first page."

Mathematicians even have problems understanding each other. To quote Cornell professor Steven Strogatz,[3]

[3] https://twitter.com/stevenstrogatz/status/1088244365724737542

> [A]lmost nobody understands each other in math seminars and colloquia. In math, we have a completely dysfunctional communication culture. Worse yet, almost all of us accept this absurd state of affairs!

So it's simply not very realistic to expect that any "outsiders" like us understand what's going on.

You might be wondering how things did come to be this way? Were they always this way?

Surely not. A few centuries ago, almost all developments in physics and mathematics went hand in hand. For example, Isaac Newton invented calculus because he needed a tool to describe mechanics.

Regarding the question of how this separation between mathematicians and other disciplines happened, once more, a great explanation is offered by Vladimir Arnold:[4]

[4] Boris Khesin. *Arnold : swimming against the tide*. American Mathematical Society, Providence, Rhode Island, 2014. ISBN 9781470416997

> "In the last 30 years, the prestige of mathematics has declined in all countries. I think that mathematicians are partially

> to be blamed as well — foremost, Hilbert and Bourbaki — the ones who proclaimed that the goal of their science was investigation of all corollaries of arbitrary systems of axioms."

He elaborates further on the contribution of David Hilbert:[5]

> "At the beginning of this century a self-destructive democratic principle was advanced in mathematics (especially by Hilbert), according to which all axiom systems have equal right to be analyzed, and the value of a mathematical achievement is determined, not by its significance and usefulness as in other sciences, but by its difficulty alone, as in mountaineering. This principle quickly led mathematicians to break from physics and to separate from all other sciences. In the eyes of all normal people, they were transformed into a sinister priestly caste . . . Bizarre questions like Fermat's problem or problems on sums of prime numbers were elevated to supposedly central problems of mathematics."

[5] V. I. Arnold. Opinion. *The Mathematical Intelligencer*, 17 (3):6–10, Jan 1995. ISSN 0343-6993. DOI: 10.1007/BF03024363. URL `https://doi.org/10.1007/BF03024363`

As a result, many mathematicians only care about mathematics and not, for example, if what they are doing is useful for other fields. Moreover, modern mathematics is mainly characterized by rigor and formality. A major contributing factor to this development was the infamous Bourbaki group and their Bourbakisme' style of mathematical literature.[6] One project of this group of mathematicians was to write textbooks that provide completely rigorous expositions. They believed that a series of such textbooks would be essential for the future progress in mathematics. While their books turned out to be influential, they made it harder for mathematicians to communicate with researchers in other disciplines like physics. To this day, many mathematicians write hard-to-read "Bourbaki style" papers. However, quite telling is that while the Bourbaki textbooks do not contain examples and non-rigorous arguments, their personal communications are full of them.

[6] This is nicely summarized by Peter Woit in his book "Not Even Wrong"

Peter Woit. *Not even wrong : the failure of string theory and the continuing challenge to unify the laws of physics*. Jonathan Cape, London, 2006. ISBN 9780224076050

This reminds me of the story recounted by Anthony Zee in his book "Quantum Field Theory in a Nutshell"[7] that a Fields Medalists[8] once told him that "top mathematicians secretly think like physicists and after they work out the broad outline of a proof they then dress it up with epsilons and deltas."

[7] A Zee. *Quantum field theory in a nutshell*. Princeton University Press, Princeton, N.J, 2010. ISBN 9780691140346

[8] The Fields Medal is in a way the Nobel Prize of mathematics.

With that said, I should emphasize that there are many great math books. It is, however, sometimes quite hard to find them. As with physics books, it usually takes some time (but is totally worth the effort) until you find a book that speaks a language you can understand. To get you started, here's a list of excellent and very readable math books:

- Calculus Made Easy by Silvanus P. Thompson,
- Math, Better Explained and Calculus, Better Explained by Kalid Azad
- The Calculus Lifesaver by Adrian Banner
- Visual Complex Analysis by Tristan Needham,
- Advanced Calculus by James J. Callahan,
- Grad, Div, Curl and all that by Harry M. Schey,
- Naive Lie Theory by John Stillwell,
- Galois' Dream by Michio Kuga,
- Visual Group Theory by Nathan Carter,
- The Shape of Space by Jeffrey R. Weeks.

These books demonstrate nicely that even quite abstract concepts can actually be understood and there is no need to wait until you get used to them.

17

History and the Big Picture

To truly understand a physical theory or model, you need to know its context. On the one hand, you should always know how a certain theory or model fits into the larger picture and how it relates to all other theories and models.[1] A second invaluable perspective is the historical context in which a theory was invented.

[1] This was already discussed in Chapter 6.

Many things are confusing if you try to understand them from a purely modern perspective. It happens quite often that certain concepts and specific formulations are overemphasized for historical reasons. The issues physicists care about change over time, but there is always a certain amount of historical baggage in any physical theory. In addition, it's often helpful to know how the inventors of a theory thought about things and what people thought before a certain discovery. And sometimes, a historical narrative simply helps to memorize certain connections and concepts.

Two amazing tomes on the history of physics are

- A Cultural History of Physics by Karoly Simonyi [2], and
- Theoretical Concepts in Physics by Malcolm Longair [3].

Moreover, there are many brilliant books that focus on the history of specific theories or individual biographies. I particularly enjoyed

- The Infinity Puzzle by Frank Close [4], which focuses on the history of quantum field theory,
- Albert Einstein: Creator and Rebel by Banesh Hoffmann[5], which helps to understand why Einstein's discoveries are so exciting,
- The Strangest Man: The Hidden Life of Paul Dirac, Quantum Genius by Graham Farmelo[6].

One thing, however, you should always remember if you study the history of physics is that confirmation bias can be quite dangerous if you try to apply its lessons to the future of physics.[7] Nima Arkani-Hamed once summarized this nicely[8]

> "That's the difficulty with the history, especially in physics. You can choose episodes from history to illustrate any polemical point you would like to make. And we never know at any time which lesson we should take. Sometimes even lessons we never learned before."

[2] Karoly Simonyi. *A cultural history of physics.* CRC Press, Boca Raton, Fla, 2012. ISBN 9781568813295

[3] Malcolm Longair. *Theoretical concepts in physics : an alternative view of theoretical reasoning in physics.* Cambridge University Press, Cambridge, U.K. New York, 2003. ISBN 978-0521528788

[4] F. E. Close. *The infinity puzzle : quantum field theory and the hunt for an orderly universe.* Basic Books, New York, 2011. ISBN 9780465063826

[5] Banesh Hoffmann. *Albert Einstein, creator and rebel.* Viking Press, New York, 1972. ISBN 978-0670111817

[6] Graham Farmelo. *The strangest man : the hidden life of Paul Dirac, quantum genius.* Faber and Faber, London, 2009. ISBN 9780571222865

[7] Confirmation bias was already mentioned in Section 15.1.

[8] The quote is from his talk "Where in the World are SUSY and WIMPS?" which is available, for example, at Youtube: `https://www.youtube.com/watch?v=dKVXxcbJ4YY`.

Regarding a better understanding of the big picture and how a certain theory or model fits into the overall narrative of physics, there are, unfortunately, currently not many books you can consult. A quite difficult book which is, however, full of beautiful insights is,

▷ The Road to Reality by Roger Penrose [9].

Alternatively, you can try,

▷ A Unified Grand Tour of Theoretical Physics by Ian D. Lawrie [10].

Moreover, I wrote two big-picture textbooks myself:

▷ Physics from Symmetry [11], in which I try to explain all theories of modern physics using symmetry principles, and

▷ Physics from Finance [12], in which I use the ideas of gauge theory to untangle the connections between the best theories of nature that we have.

[9] Roger Penrose. *The road to reality : a complete guide to the laws of the universe*. A.A. Knopf, New York, 2005. ISBN 978-0679454434

[10] Ian Lawrie. *A unified grand tour of theoretical physics*. CRC Press, Taylor & Francis Group, Boca Raton, 2013. ISBN 978-1439884461

[11] Jakob Schwichtenberg. *Physics from symmetry*. Springer, Cham, Switzerland, 2018a. ISBN 978-3-319-66631-0

[12] Jakob Schwichtenberg. *Physics from finance*. No-Nonsense Books, Karlsruhe, Germany, 2019d. ISBN 978-1795882415

18

Philosophy

The relationship between physics and philosophy is currently far from intact. This is perfectly exemplified by the fact that philosophy has become a derogatory term in physics circles. "Are you a philosopher or what?" is used as an insult for someone who asks too many "why" questions. Moreover, there is not a single lecture on the philosophy of physic or science in a normal physics curriculum.

"So what?", you might argue. Maybe philosophy has simply become obsolete.

After all, many famous physicist like, for instance, Neil de Grasse Tyson[1], Lawrence Krauss[2] and Steven Hawking[3] made public statements along these lines.

In my humble opinion, these statements are completely missing the point. Of course, it makes sense to criticize certain areas of philosophy that have become obsolete. But philosophy as a whole still has a lot to offer. Moreover, phi-

[1] https://www.youtube.com/watch?v=ltbADstPdek

[2] https://www.theatlantic.com/technology/archive/2012/04/has-physics-made-philosophy-and-religion-obsolete/256203/

[3] https://www.theguardian.com/commentisfree/belief/2010/sep/08/stephen-hawking-philosophy-maths

losophy is simply not competing with science but instead, can offer invaluable guidance for scientists.

Einstein articulated this point of view perfectly:[4]

[4] Einstein letter to Robert Thornton, Dec. 1944.

> "So many people today – and even professional scientists – seem to me like somebody who has seen thousands of trees but has never seen a forest. A knowledge of the historic and philosophical background gives that kind of independence from prejudices of his generation from which most scientists are suffering. This independence created by philosophical insight is — in my opinion — the mark of distinction between a mere artisan or specialist and a real seeker after truth."

So as before, if your only goal is to make a career in physics, you can probably safely ignore philosophy. But if you want to get a truly deep grasp of reality, philosophy will prove to be a constant and trustworthy companion.

In fact, we currently live in times in which physics needs help from philosophy more desperately than ever before. Nobel Prize winner famously Sheldon Lee Glashow noted[5]:

[5] Tian Cao. *Conceptual foundations of quantum field theory*. Cambridge University Press, New York, 1999. ISBN 9780511470813

> "[E]verybody would agree that we have right now the standard theory, and most physicists feel that we are stuck with it for the time being. We're really at a plateau, and in a sense, it really is a time for people like you, philosophers, to contemplate not where we're going, because we don't really know and you hear all kinds of strange views, but where we are. And maybe the time has come for you to tell us where we are. 'Cause it hasn't changed in the last 15 years, you can sit back and, you know, think about where we are."

Fundamental physics is in a decade-long phase of stagnation and so many speculative ideas have failed to live up to the hype that people start to question everything. Maybe we should give up and declare that everything is simply anthropic?[6] But wait, this isn't a valid scientific hypothesis since how would you ever test it? Now that we are talking, how should we deal with theories like string theory that do not make any definite predictions? Are they part of science after all? And when do we stop investigating a certain hypothesis like supersymmetry that makes predictions, if it's always possible to modify the corresponding models a little bit to comply with the latest data? And does it really make sense to continue our search for further elementary particles or are they simply a modern incarnation of the infamous epicycles? Regardless of your stance on such questions, if you ponder them, you're doing philosophy.

[6] The anthropic principle states that certain things in nature are the way they are simply because otherwise we couldn't be here observing them.

In other words, philosophy can help to decide which questions are worth pursuing. After all, "the hardest part of research is to decide exactly what question you're going to work on.", as Fields Medalist Edward Witten once noted[7],

[7] David Appell. When supergravity was born. *Physics World*, 25(09):32–36, sep 2012. DOI: 10.1088/2058-7058/25/09/36

To get some feeling for what the philosophy of science can offer, try

▷ Mark A. Notturno's summary of Karl Popper's thoughts on the scientific method.[8],

[8] `http://www.the-rathouse.com/Intro-Philos-Sci/2-Method.html`

▷ Upgrade your cargo cult for the win by David Chapman[9],

[9] `https://meaningness.com/metablog/upgrade-your-cargo-cult`

▷ Science: Key Concepts in Philosophy by Steven French[10],

[10] [French, 2016]

▷ Theory and Reality by Peter Godfrey-Smith[11].

[11] [Godfrey-Smith, 2003]

Fantastic books to dive deeper are:

- ▷ Against Method by Paul Feyerabend[12],
- ▷ Every Thing Must Go by James Ladyman and Don Ross.

[12] Paul Feyerabend. *Against method.* Verso, London New York, 2010. ISBN 978-1844674428

In addition, there is not just the philosophy of science in general but also the philosophy of physics in particular. Physical theories, models, formulations and interpretations can raise many interesting specific philosophical questions. Moreover, if you dive really deep into any physics topic, you will notice at some point that many insightful discussions can be found in papers by philosophers.

Great examples are the role of the renormalization group and emergence[13], the puzzle of why gauge symmetries are so important even though they only represent redundancies in our description[14] and the question of why the Higgs mechanism works even though spontaneously breaking local symmetries is impossible[15].

[13] A fantastic book on this subject written by a philosopher is "The Devil in the Details" by Battermann [Batterman, 2002]

[14] I have written a somewhat philosophical paper on the subject .

Jakob Schwichtenberg. Demystifying Gauge Symmetry. 2019a

[15] See, for example, http://jakobschwichtenberg.com/higgs-intuitively/.

The most famous example, however, is the question of how we should interpret quantum mechanics. There are dozens of different proposals and since they are only different interpretations of the same mathematical machinery they lead to the same experimental predictions. Thus, it's really a matter of philosophy which interpretation you prefer.

Often (young) researchers are actively discouraged to spend time on topics likes this since many consider it a waste of time.[16] You might even argue that this makes sense. The formulas work and as long as their predictions agree with what we observe in experiments, why should we care about their deeper meaning?

[16] For a nice essay on this topic, see "*Shut up and let me think. Or why you should work on the foundations of quantum mechanics as much as you please*" by Pablo Echenique-Robba; https://arxiv.org/abs/1308.5619

Well, first of all it is always worth remembering that "our aim in physics [is] not just to describe nature, but to explain nature", as noted by Steven Weinberg[17].

[17] Tian Cao. *Conceptual foundations of quantum field theory*. Cambridge University Press, New York, 1999. ISBN 9780511470813

Secondly, physics is far from finished and as long as there are open questions, we need to care about the meaning of the formulas that we handle. This becomes especially obvious if we consider a concrete problem like quantum gravity. Despite decades of work by many of the best physicists there is still no experimentally verified, consistent quantum theory of gravity.[18] This is extremely puzzling because both, quantum field theory and general relativity, seem to work perfectly as far as we know. But if we try to combine them, we get something seemingly inconsistent.

[18] I'm not an expert when it comes to quantum gravity and I thus try to formulate everything carefully. Be warned that since, so far, there is no experimental test that tells us which proposed model of quantum gravity is the right one, many researchers are quite dogmatic about their preferred approach.

Since physicist struggle with this problem for decades now, it may seem worthwhile to take a step back and critically examine all of our assumptions and even the goal that we are trying to achieve. For example, maybe it's not gravity that needs to be modified to fit into the quantum framework, but the quantum framework itself? This would imply that the quantum rules we know so far are not truly fundamental and thus we are right in the middle of the interpretation-of-quantum-mechanics business.

Two great books that discusses philosophical aspects of quantum mechanics in detail are,

- ▷ Interpreting Quantum Theories by Laura Ruetsche [19] and
- ▷ The Interpretation of quantum mechanics by Roland Omnes[20].

[19] Laura Ruetsche. *Interpreting quantum theories : the art of the possible*. Oxford University Press, Oxford, 2013. ISBN 978-0199681068

[20] Roland Omnes. *The interpretation of quantum mechanics*. Princeton University Press, Princeton, N.J, 1994. ISBN 978-0691036694

As usual, the best advice I can give you is to give it a try. Don't trust arguments from authority but also don't trust me. Instead, read one or two of the really great texts on the philosophy of science and then decide for yourself whether or not you find the ideas useful.

19

Sociology

A final aspect of physics that is too often ignored is its human dimension.[1] To an outsider it might be shocking to hear that physicist are not solely driven by their search for truth. But physicists are humans like anyone else and no one is completely immune to the thoughts, words and actions of their peers. There is a global physics community and many decisions and beliefs within it are heavily influenced by sociological forces.

[1] We touched already on some aspects of this topic in Section 15.1, in which we discussed common fallacies and biases.

For example, Howard Georgi once summarized an important but very humane aspect of theoretical physics research[2]:

> During periods, without experiment to excite them, theorists tend to relax back into their ground states, each doing, whatever comes most naturally. As a result, since different theorists have different skills, the field tends to fragment into little subfields. Finally, when the crucial ideas or the crucial experiments come along and the field regains its vitality, most theorists find that they have been doing irrelevant things.

[2] P. C. W. Davies. *The New physics*. Cambridge University Press, Cambridge Cambridgeshire New York, 1989. ISBN 9780521304207

Therefore, to interpret claims and statements that are made by physicist properly, it can certainly help to study their sociological context.[3]

[3] Of course, just as with cognitive biases, knowing about sociological forces (a.k.a. social biases) does not make you immune to them. But it certainly helps to withstand them.

Excellent short reads on sociological aspects of modern physics are

- ▷ What Does Any of This Have To Do with Physics? by Bob Henderson[4], and
- ▷ Mimetic traps by Brian Timar[5].

[4] `http://nautil.us/issue/43/heroes/what-does-any-of-this-have-to-do-with-physics`

[5] `https://www.briantimar.com/notes/mimetic/mimetic`

Great books that deal with various sociological aspects of the physics community in more detail are:

- ▷ My Life as a Quant by Emanuel Derman [6],
- ▷ The Trouble With Physics by Lee Smolin [7],
- ▷ Not Even Wrong by Peter Woit [8],
- ▷ Lost in Math by Sabine Hossenfelder [9].

[6] Emanuel Derman. *My life as a quant : reflections on physics and finance*. Wiley, Hoboken, N.J, 2004. ISBN 9780470192733

[7] Lee Smolin. *The trouble with physics : the rise of string theory, the fall of a science and what comes next*. Penguin, London, 2008. ISBN 9780141018355

[8] Peter Woit. *Not even wrong : the failure of string theory and the continuing challenge to unify the laws of physics*. Jonathan Cape, London, 2006. ISBN 9780224076050

[9] Sabine Hossenfelder. *Lost in math : how beauty leads physics astray*. Basic Books, New York, NY, 2018. ISBN 9780465094257

Next, after discussing all these topics that might be worth spending some time on, we should finally start talking about how you should go about studying them. That's what the next part is about.

Part V
Advice and Principles

"Listen, then make up your own mind."

Gay Talese

If you want to develop a truly deep understanding of physics, you need a solid plan since as the popular saying goes, "If you fail to plan, you plan to fail." Passion and hard work can take you quite far, but directing that passion and hard work in an efficient manner over time is what leads to true expertise.

At the beginning of Part III, we already talked about roadmaps and the importance of creating your own. In addition to such a general roadmap that defines in which order you want to tackle various theories and concepts, you need a more detailed plan that outlines how you want to learn a specific topic.

Most people don't even spend five minutes to reflect how they want to understand something, as Richard Feynman observed in his book Surely you're joking, Mr. Feynman:

> "I don't know what's the matter with people: they don't learn by understanding; they learn by some other way — by rote or something. Their knowledge is so fragile!"

The difference between fragile and antifragile[10], understanding is nicely illustrated by the following apocryphal story.

Nobel Laureate Max Planck had a chauffeur who drove him to all the public lectures he was invited to. After a while, the chauffeur had memorized Planck's standard talk and asked him "Professor Planck, maybe we can switch places?" Planck agreed and the chauffeur gave the lecture. Everything went fine until at the end of the lecture, one of the attending professor stood up and asked an extremely difficult question. But the chauffeur kept his calm and said "I'm quite surprised that someone at the prestigious university of

[10] Antifragile is a term coined by Nassim Nicholas Taleb in his brilliant book of the same name [Taleb, 2012]. It describes the opposite of fragile and is not the same as robust. Something robust only doesn't break when damaged, while something antifragile actually gets stronger. Therefore, in the context of knowledge, developing an antifragile understanding means that it actually deepens when certain assumptions or conclusions are proven to be wrong. This is, of course, exactly what we need as physicists. We will talk about this in more detail in Chapter 24.

Munich is asking such an elementary question, so I will ask my chauffeur to answer it".

In this story, Max Planck has truly deep knowledge, while the chauffeur's knowledge is extremely fragile. The chauffeur might be able to give the talk and make a great impression, but he certainly isn't able to go anywhere beyond that. He is entirely confined by the phrases he memorized. Of course, he could also learn how to answer questions that are asked after the talk. But as soon as some facts change and the talk needs to be fundamentally modified, he would be completely lost. If you're building a card house, even the slightest disturbance can bring the whole structure crashing down. In contrast, for Max Planck the talk only represents the tip of the iceberg, and he will have no problem giving a modified talk if new information becomes available. Most importantly, we can imagine that Max Planck in our story would be delighted if something new is discovered, even if it invalidated certain some of his long-held assumptions. New discoveries represent an opportunity to deepen his knowledge. He will be able to understand the wide-ranging consequences and how the various parts of physics are affected. In other words, his knowledge will increase through disturbances. This is the hallmark of antrifragile understanding. In contrast, we can equally imagine that the driver would be quite annoyed by new discoveries since many if not all facts and phrases he memorized would become obsolete. His fragile knowledge is in constant danger of becoming useless.

The message to take away is that without a plan that helps you to develop an antifragile understanding, you will struggle for years and in end, still understand things only superficially.

Now, how could such a plan look like?

Here's what I usually try to do.

First of all, I try to find inspiration, motivation and orientation. If I don't care about a subject, I will never be driven enough to develop a really deep understanding. Rough passages are sometimes inevitable and without proper intrinsic motivation, I will always be tempted to simply accept a superficial level of understanding. Moreover, without proper orientation, I will quickly get lost in technical details. Thus, it's always a good idea to get oriented by reading some less technical book first. This will give me some awareness of the big picture and what needs to be learned.[11]

[11] We talked about big picture books in Chapter 17.

Afterwards, I pick one textbook and read it quickly from cover to cover. In this first reading, I skip all technical details and ignore all exercises. Otherwise, it's too easy for me to get frustrated and lose the forest for the trees. I spent a lot of time choosing the right first book since, as with people, the first impression a subject makes on you is critical.

One key here is to understand that usually when you struggle, it's not your fault. Far too many people waste weeks if not months trying to understand explanations that were simply not written to be understood. After some time they then conclude, "well, things are complicated", memorize a few key facts and move on. This is how fragile knowledge is born. [12]

[12] We will talk about this in more detail in Chapter 20.

Thus, my first goal in learning any subject is always to find one book that speaks a language I can easily understand.

But even if you find a great book to start with, there are still many traps. For instance, if you spend all of your time reading, your understanding will always remain fragile. In particular, you will never know what you do not know. Instead, you need to actively engage with what you're reading and make a conscious effort to identify holes in your understanding. You constantly need to ask questions and then try to find answers wherever you might find them.

Thus, here's what I typically do after reading a first book from cover to cover:

First of all, I try to identify the fundamentals since otherwise I would quickly feel overwhelmed. In the beginning it always seems as if I had to memorize thousands of new facts, concepts and ideas. But this is an illusion. There are always just a few core principle that govern the field and thus my first job is to identify them.

Once I've identified and mastered the fundamental, my understanding will be quite robust.

As soon as I've identified the fundamentals, I try to draw a map that encodes the connections between the various concepts and equations.[13] Afterwards, I try to teach the subject to a fellow beginner. This is the most effective method to identify gaps in my understanding. During this process of teaching, I quickly discover many things that I do not fully understand. I then try to close these gaps by reading various chapters and sections in lots of additional textbooks or consult different sources. This way I will not only develop a deep understanding but also a unique perspective on the subject. We will discuss all of this in more detail in Chapter 21.

[13] You can find several examples of such maps in the subject guides in Part III.

Finally, since my goal is usually to use what I learned in some way or another, I practice by solving problems. This means that in this final step, I try to learn how I can apply the methods and equations to describe concrete systems. The role of practice is discussed in more detail in Chapter 22.

In summary:

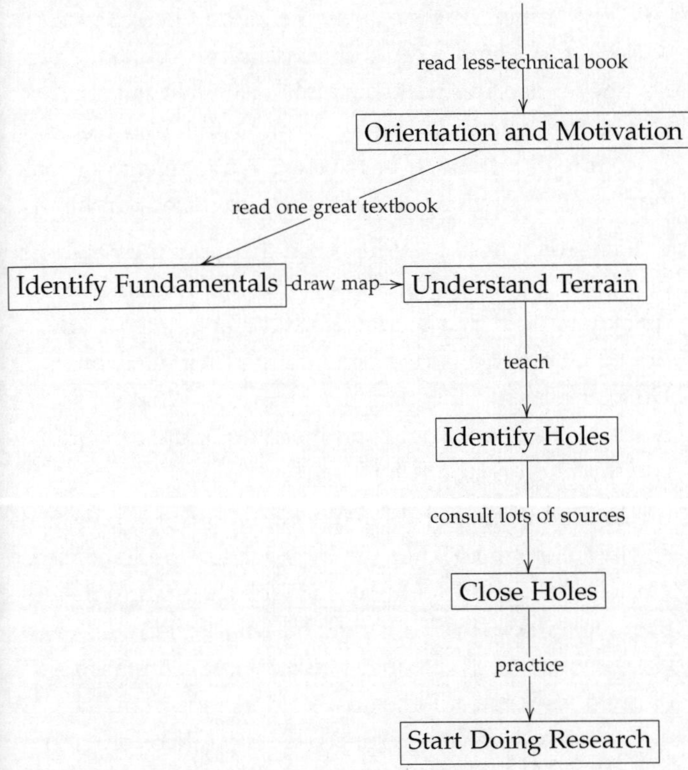

As an example, here's roughly how I learned quantum field theory

▷ I got interested in quantum field theory and developed some kind of big picture overview by reading QED: The Strange Theory of Light and Matter" by Richard P. Feynman.

▷ Afterwards, I skimmed through dozens of textbooks like An Introduction To Quantum Field Theory by Peskin and Schröder, Quantum Field Theory in a Nutshell by Zee, and The Quantum Theory of Fields by Weinberg. None of them made much sense to me.

▷ Then I discovered Student Friendly Quantum Field Theory by Klauber. It was exactly the kind of book I was looking for. Klauber carefully explains all subtle details and never declares that some things are just too obvious to waste time explaining them. Within four weeks or so, I read it from cover to cover. While reading, I took hundreds of small notes and drew dozens of mindmaps.

▷ Afterwards, I tried to write down the fundamental ideas and their connections in my own words. In doing so, I quickly noticed that there were still many holes in my understanding. Where exactly do the Klein-Gordon, Dirac and Maxwell equation come from? What's a spinor really? What's the origin and meaning of the canonical commutation relations for quantum fields?

▷ I then tried to answer these questions by consulting many additional textbooks like the Introduction to Elementary Particles by Griffiths, Quantum Field Theory and the Standard Model by Schwartz, and Quarks, Leptons and Gauge Fields by K. Huang. I quickly learned that most of my questions could be answered using a branch of mathematics known as group theory.[14] I therefore also went down this rabbit hole and read various chapters in books like Naive Lie Theory by Stilwell and An Introduction to Tensors and Group Theory for Physicists by Jeevanjee. Each time I learned something new, I expanded my notes and published some of them on my website like, for example, `http://jakobschwichtenberg.com/naive-introduction-lie-theory/` or `http://jakobschwichtenberg.com/adjoint-representation/`.[15]

▷ To refine my understanding further, I started solving basic exercises like the calculation of the tree level muon decay amplitude. In addition, I started reading research papers on grand unified models[16] and tried to replicate

[14] Group theory is what we use to describe symmetries.

[15] A polished version of my notes is now published as a book titled "Physics from Symmetry" [Schwichtenberg, 2018a].

[16] Grand unified models live exactly at the intersection of my interests: quantum field theory and symmetries.

their results.

> This journey is far from finished. I'm still learning new things about quantum field theory and I still publish new notes online, for example, `http://jakobschwichtenberg.com/understanding-symmetry-breaking-goldstones-theorem-intuitively/`.

This is roughly how I try to tackle any new subject.

However, you should primarily use my guidelines and plan as a source of inspiration. You need to adapt it to your own needs since even the best plan is worthless if you cannot stick to it. Learning physics is a journey which must be continued consistently for quite some time before things start to fall into place. Therefore, if the plan you're following is not fun, you're doomed to fail. Individuality is key to long-term success.

For this reason, I've purposefully not included timelines that outline how long it should take to learn, say, classical mechanics. As mentioned in previous chapters, everyone has a different background, different interest, and different resources available. Thus, one student may only want to spend one month learning classical mechanics, while another student may want to spend a whole year since he has never heard of Newton's laws before.

The best advice I can give you is to start with a plan that looks roughly like the one given above. As you follow the plan, you should observe what works for you personally and always adjust your plan accordingly.

To use a cooking analogy, it's okay to start with recipes but ultimately you need to learn how to taste to create a perfect menu using the resources you have available.

In addition, there are two further observations that may help you in your journey. The first one is that learning physics takes time for everyone and is never a linear process. You need to revisit topics over and over again until you've truly internalized them. This is discussed in Chapter 23. Moreover, it's easy to get fooled by the illusion of simplicity or the illusion of complexity, depending on what stage in your learning journey you're currently at. In Chapter 24, we will thus discuss the concept of informed simplicity and further important aspects of learning and understanding.

With this rough outline in mind, let's dive right in.

20

One Thing You Must Understand

As a beginner student it's easy to feel overwhelmed and stupid. We are usually trained to consider an inability to understand something as a failure on our part. So it took me quite some time to realize that the real problem is that most textbooks aren't written for the reader, but for the author. Similarly, most lectures are not designed to help the audience understand.

My realization started with a little sentence in a book on classical mechanic by Cornelius Lanczos[1]

[1] Cornelius Lanczos. *The variational principles of mechanics*. Dover Publications, New York, 1986. ISBN 9780486650678

> "Many of the scientific treatises of today are formulated in a half-mystical language, as though to impress the reader with the uncomfortable feeling that he is in the permanent presence of a superman."

Physics, like any scientific discipline, is to some extent also about prestige. Writing a super-complicated treatise may help the author get recognized as super-smart, but surely isn't very helpful for the reader. Many author overcomplicate because they are afraid that an obvious, simple approach wouldn't seem "serious" enough.

Moreover, the probability that someone criticizes your treatment is much higher for a book that every reader understands. Every understandable sentence in a book is risky.

It's much safer to write down equation after equation without talking much about them. The equations can be checked straight-forwardly and there is little room for criticism. But as soon as an author starts writing about the meaning of the equations, the why and the how, things become dangerous. Sentences that interpret things and put them into context are often immensely valuable, but at the same time, are in constant danger of being criticized.

There is just no way to make them as bullet-proof as an equation. That's why quite often the authors' ego wins the fight and explanatory sentences are kept to a minimum.

This observation was nicely summarized by Michael Lewis in his book Undoing Project [2]:

[2] Michael Lewis. *The undoing project : a friendship that changed our minds*. W.W. Norton & Company, New York, 2017. ISBN 9780241254738

> "[Academic authors] aren't trying to engage their readers, much less give them pleasure. They're trying to survive them."

Formulated differently, many academics write terribly because they write for an audience out to destroy them. They are terrified to make a mistake. The same is true for professors in lectures. Only a tiny minority of all professors and teachers are brave enough to teach without any fear of damage to their self-image.

I still remember how afraid I was before I published my first book. Once a book is published, there is no point of return. Everyone in the world can read your thoughts and laugh about them. Thus, I felt an incredibly strong urge to make my book more bullet-proof. I wanted to delete and rewrite hundreds of sentences. Luckily, I had a deadline and ultimately many somewhat unconventional statements (that readers now find extremely helpful) survived the purge.

So, always keep in mind that books that are hard to understand are written with the author's ego in mind and not the readers' needs.

It's the authors job to make it easy on the reader. If you don't understand something, it's the authors fault, not yours. If you are reading a book that you find hard to understand, put it down and pick up another book. This is quite hard for most people because we are taught in school that books are something you have to finish. So you have to actively try to get into the habit of quitting books you don't like. To quote science fiction author and autodidact Jorge Luis Borges:[3]

> "If a book bores you, leave it; don't read it because it is famous, don't read it because it is modern, don't read a book because it is old.... If a book is tedious to you, don't read it; that book was not written for you."

This is especially important because every hour you spend reading a boring book is an hour you could've spent reading a great book. Of course, the word "great" has no absolute meaning here. Sometimes, you are just not ready for a particular book.

[3] Jorge Borges. *Professor Borges : a course on English literature*. New Directions Books, New York, 2013. ISBN 9780811218757

So please don't let yourself get demotivated! You simple need to find a book that you can easily understand. Physics is never really complicated, only badly explained.

To find great books, you have to skim quickly through all the books that spark your interest.[4] Then, once you've found one that you truly like, give it all your attention and devour it.

[4] You can skim through books offline in a library or online using Google Books.

You are probably wondering why people read and recommend all these complicated books at all?

Again, ego is the answer. Most "standard books" and textbooks recommended by professors aren't particular well-suited for students. These books don't have the best explanations, but are recommended because they are safe to recommend. There is little room for criticizing them because the explanations and illuminating remarks are kept to a safe minimum.

In addition, recommending a book that is hard to understand is good for your ego. Indirectly you're signalling this way: "Well, I understand and even enjoy this super-complicated stuff. I'm smart!"

A recommendation for a book that explains everything in great detail signals: "I needed such dummy treatment to understand the subject. I'm not particular smart."

So, don't panic, don't feel stupid. It's the same for everyone. Most of the time the recommendations of books that are hard to understand happen with no ill intent. Your professor learned the subject some decades ago, so he's not the best person to get recommendations from. Instead, try to find out what books other beginners are truly excited about.

Be assured that if you read some treatment of the subject that was written with the readers needs in mind first and only afterwards read the super-complicated stuff, you will understand it too.

The message to take away is: stop reading books you don't understand immediately! There is absolutely no reason to feel ashamed or stupid. If you don't understand something, simply search for another explanation that makes sense to you! Or, in the words of the author Austin Kleon:[5]

[5] https://austinkleon.com/2019/03/21/how-to-read-more-3/

> "Stop reading what you think you should be reading and just read what you genuinely want to read."

Moreover, on a more microscopic level it is important to recognize that it's not just acceptable and smart to quit books, it's also often reasonable to skip certain paragraphs or sections in a book. Whenever a book gets boring, I skip ahead and only continue once I've found a paragraph that sounds interesting. I feel no obligation to read every single sentence of a book. In physics textbooks in particular, it often makes sense to skip all long calculations at first encounter and revisit them later once you have some solid understanding of the big picture and of the importance of the calculation. Far too many students get stuck because they don't understand some detail of a calculation which isn't really that important in the grand scheme of things.

The practical lesson to be learned here is that if you want

to learn a subject, you need to spend whatever time is necessary to find a textbook which speaks a language that *you* can understand easily. While reading the book, you should feel that the author really cares about you and does everything to help you understand. Then read this textbook from cover to cover. This is how you get a foot in the door. In the following chapter, we will discuss how you can level up afterwards.

21

The Master Key

Reading is easy. Thinking is hard. As discussed in the previous chapter, it makes sense to start by reading one great book. But reading is only the first step. To internalize whatever you're trying to understand right now, you need to question everything actively and ponder what you read.

A great guideline is Feynman's famous mantra:

> "What I cannot create, I do not understand."

Only if you're able to start from nothing and re-derive everything as needed, you can truly claim to understand it.

And it's not sufficient to just memorize the steps someone else told you about. You not only need to know the steps, but you must *understand* them. Being able to recite a bunch of definition simply does not equate understanding. It's not a valid approach to justify a specific step or a whole path with the argument that authority X says so.[1] You need to verify everything yourself. Otherwise, you will never be truly confident in what you understand and what you don't. *If* you adopt something someone else said, it's only after you have verified it for yourself.

[1] We talked about arguments from authority in Chapter 15.1.

You can find a great example of this in Volume 1 of the Feynman Lectures on Physics.[2] In just four pages he takes you from counting numbers on the hand to calculus. It's obvious that he understood numbers at such a deep level that he didn't have to memorize anything.[3]

[2] You can read the relevant chapter here: `http://www.feynmanlectures.caltech.edu/I_22.html`.

[3] A much longer example is my book "Physics from Symmetry" in which I attempt to derive all theories of modern physics solely using symmetry principles.

Arguably the most important aspect of all of this is that you need to develop a habit of radical intellectual honesty. An intellectual honest person says "I don't know" all the time — not just to others but also to themselves. You need to resist the urge to act smart and to move on prematurely. It is essential to admit the limits of your understanding because it's the only way to make progress.

Moreover, you need to be able to walk the whole path in your own shoes. Forget what anyone else says. Just figure out what's true.

Surprisingly, many people disagree with this statement. They argue that the path is just too long and no one knows all the steps.

What these people don't realize is that there is never just one road from A to B. Sure, the path they were shown during their guided bus tour might have been incredibly confusing. But there are other roads that are not only more beautiful but also astonishingly simple and short. In addition, they don't realize that it's easy to know a path by heart, if you've discovered it yourself. So if you're someone who has never discovered a path of his own and only knows the standard routes, Feynman's mantra[4] may seem like an unattainable ideal. But if you're out there, backpacking through the wonderful continent that we call physics, you will understand quickly that there simply is no alternative.

[4] "What I cannot create, I do not understand."

To summarize, you need to think for yourself and figure out what's true. There is no choice about this. There are no turnkey solutions and you can't walk in someone else's shoes. This is really the master key. If you only remember one thing from this book make it this one.

You might be wondering how you can actually create a path of your own. And how can you check whether you truly know and understand all the steps from A to B?

The answer is simple: teach.

21.1 Teach

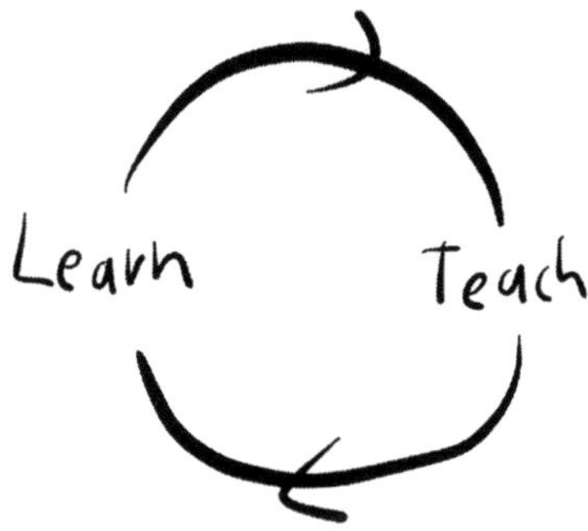

While there are numerous variants, I think the following (paraphrased) line from Kurt Vonnegut's Cat's Cradle captures the general idea perfectly:[5]

[5] Similar lines are often attributed to Richard Feynman and Albert Einstein.

> "Any scientist who can't explain to an eight-year-old what he is doing is a charlatan. "

This is not only a useful heuristic to spot imposters but also to assess your own understanding.

In particular, whenever you're unsure whether you truly understand something — which should happen all the time — try to teach it. Simply try to explain it to someone else who is not familiar with the topic in the most unadorned terms.[6]

[6] The method of learning by teaching is sometimes called the Feynman method or Feynman technique.

Of course, most of the time you won't find anyone who is willing to listen to you. Luckily, it's sufficient to explain things to an imaginary friend by writing it down.

Imagine you're sitting next to a good friend (or a younger version of yourself) and then write down how exactly you would explain things to him or her. Really, write like you would talk. Avoid all big words and try to make everything as simple as possible (but not any simpler).[7] Your goal is not to impress anyone but solely to help your friend understand. There is no big audience. It's just the two of you. Simply write what you would want to read.

[7] The aphorism: "Everything should be made as simple as possible, but no simpler," is often attributed to Albert Einstein.

And don't worry about grammar or typos or if you're really describing the best path. Just get your thoughts out of your head and onto the paper or screen. Then later, revisit what you wrote and refine your notes over and over again.

Writing down your thoughts is the most efficient way to spot holes in your understanding. Quite often when I start to write ideas down I find that they don't make much sense once I see them spelled out on my screen. And any idea that doesn't stand the test of being written down, wasn't good enough in the first place. By writing things down you can test whether your thoughts really make sense. Transferring vague feelings into words leads to clarity and surprising insights. I'm constantly amazed at how much I learn by writing things down, even if no one but me reads them.

In addition, your writings will be an invaluable support structure as you move further. We forget stuff all the time and there is no better way to relearn something than by reading your own explanation. Moreover, even the best ideas start to fade away after some time if you don't write them down. So it's always worth the effort to save them in the form of written words.

But this method only works if you do not simply copy paragraphs from some textbook. While you certainly don't need to reinvent the wheel, you should attempt to rearrange existing puzzle pieces in novel ways and add your own flavor. This is how you build your own paths.

A great way to get started is to summarize the first book you read on a topic. For example, if your goal is to learn quantum mechanics, you could try to answer the following questions:

- ▷ Why should you care about quantum mechanics? Why is it necessary?
- ▷ How can we describe quantum objects? What are the basic assumptions, concepts and equations?
- ▷ What's the meaning of the Schrödinger equation? Is it similar to any familiar equation?
- ▷ What's the meaning of solutions of the Schrödinger equation?
- ▷ ...
- ▷ How is quantum mechanics related to classical mechanics and quantum field theory?

Then try to write down your own answers to these questions. While doing this, constantly ask yourself:

- ▷ How do I know?
- ▷ Says who?
- ▷ How specifically?
- ▷ Does it always happen?
- ▷ Can I think of an exception?
- ▷ What is stopping me from doing ...?

Don't try to plan in detail what you will write about. Use your list of questions as rough outline and then get started. The rest will take care of itself since you will have most of your best ideas while you write and not in advance.

You will get your best idea from writing, not from some magical preliminary step or by looking at a blank page. It's only through the writing process itself that the fuzzy ideas we have floating around start to crystallize. Quite poetically writer David Perell observes:[8]

[8] https://twitter.com/david_perell/status/1164727319505883136

> "Your best ideas are hiding in the shadows. They'll come to light once your fingers hit the keyboard."

Of course, it's intimidating to start writing without a clear idea of what you're going to write about. But the key really is to push through this invisible wall of resistance and get started even though the ideas are still fuzzy in your head.

You should also consider publishing your notes as you go.[9] This forces you to clarify your understanding and articulate it in a way that other people will understand. Moreover, it can be a powerful source of motivation since your notes possibly can help hundreds of other students.

[9] I personally do this in the form of blog posts at www.jakobschwichtenberg.com and at www.physicstravelguide.com.

Of course, putting your work out there is scary. However, the primary source for this fear is that we were all trained in school to be afraid of mistakes. In the real world, you're not punished for a mistake in your notes or an essay or if you admit that you don't know something. Instead, if someone points out a mistake, it's an opportunity for you to refine your understanding.

The key is to recognize that almost everything you produce will fall embarrassingly short of your ambition. Embrace it and continue to make improvements along the way where you can.

Just always be honest when you don't fully understand something or aren't sure that a statement is correct. The feedback that you are then able to receive can accelerate your learning journey dramatically. In addition, publishing your notes online is often the best way to connect with like-minded people. For example, John Baez describes his experiences as follows[10]

[10] http://math.ucr.edu/home/baez/mathblogs.pdf

> "My introduction to blogging came in 1993 when I started an online column called "This Week's Finds in Mathematical Physics". The idea was to write summaries of papers I'd read and explain interesting ideas. I soon discovered that, when I made mistakes, readers would kindly correct them - and when I admitted I didn't understand things, experts would appear from nowhere and help me out. Other math bloggers report similar results."

So don't wait until you understand a topic perfectly since this day will most probably never come. At the beginning your thoughts will be too scattered and you will not be able to write things down in a coherent fashion. But after some time, there will be a point where you start to see things clearly. This is the perfect point to start writing even if you still don't understand all the details. If you wait too long, however, things will start to look too obvious and you will no longer be motivated to write them down in a proper way. This has happened countless times to me and I still regret that I didn't write my ideas down earlier.

You might object that you're simply not qualified enough to teach and there are already enough explanations.

First of all, this doesn't matter since you're writing for an audience of one: yourself. You really shouldn't care what others are thinking. Otherwise ,it happens far to quickly that you become intellectually dishonest and start acting smart.

Secondly, there is no such thing as too many explanations. Different people find different perspectives valuable. If someone doesn't like how you approach things, he can simply ignore your writing. But usually, the situation really looks like this:[11]

[11] The following image is adapted from a popular cartoon by a tumblr user named stuffman. (`https://sqbr.tumblr.com/post/92103436228/the-artist-putting-a-simple-cake-next-to-a-much`). In case you can't read my handwriting, the text in the speech bubbles reads: "that guy's cake is way better than mine" and "Wohoo! Two cakes!".

I think it would be amazing if there were hundreds of different explanations of any physics concept.

Moreover, once again you should keep in mind that beginners are usually better teachers than experts.[12] In the cake analogy depicted in the cartoon above, the cake by a professional chef looks great, contains lots of fancy ingredients and will be loved by food critics. But at least, I for my part, usually prefer much simpler creations that taste more similar to what my grandmother used to bake.

[12] We discussed this already in Chapter 20. If you want to learn more, the "curse of knowledge" is what you should be searching for.

So in summary, I think it's totally worth it to get over that latent fear and start sharing what you learn since it will allow you to advance much faster and make useful connections along the way.

Now, how concretely do you develop a path of your own? Above, I've already mentioned that the key is to draw from many sources and rearrange existing ideas in novel ways. Of course, to do this, you need to be familiar with many sources.

So let's talk about how you can develop a unique perspective by familiarizing yourself with the works of many thinkers.

21.2 Jump

As discussed in Chapter 20, most books are not worth reading them cover to cover — not necessarily because they are bad, but because they are not what you need at a certain point in your journey.

Instead, after reading *one* great book to get a preliminary overview, you should focus on individual chapters in lots of different books to answer specific questions. Of course, you shouldn't restrict yourself to books. Read voraciously whatever helps you understand. An epiphany is just as likely to come from a paper on the arXiv than from a book chapter or from a blog post. You need to jump around until your question is answered and then move on to the next one.

Many people find this hard because they are taught that you *need* to finish a book before you're allowed to move on to the next one. If you stick to such a book-after-book approach you will probably never get anywhere. At least, you will dramatically slow down your progress if you dwell for too long on a bad explanation that simply wasn't written to be understood. As soon as you get the feeling that a certain explanation doesn't help you, dismiss it and move on to the next one. It is, however, important to keep in mind that even an otherwise horribly bad book can contain a certain nugget of wisdom that elevates your understanding to a completely new level. Moreover, whether you find a certain book helpful or not always crucially depends on what you've read previously. At some later stage, a book you initial thought to be useless can turn out to be an invaluable guide. So you need to revisit the same books over and over again.

At some deep level you absorb everything that you read. So

feel free to skim and jump around. If you do this with some specific question in mind, at some point, things will fall into place and it will make "click" in your head.

Jumping around between books is essential because there simply is no single book which will be able to answers all of your questions. Any author or teacher can only answer a few questions really well.

Take me as an example. I have written textbooks on topics ranging from classical mechanics to quantum field theory. But still, I certainly don't have all the answer. Student's regularly point out things I have never thought about before. There is a whole universe of things I know nothing about. And this is true for any author or teacher.

You need to learn from multiple sources because Nature is far more imaginative than any single author or teacher.

22

Practice

To truly master any topic, you need more than mere knowledge. It's also not sufficient to be able to explain your knowledge to a hypothetical eight-year-old. You also need to know how to apply it.

Of course, if you only want to learn a topic like quantum mechanics as an end in itself, that's perfectly fine. In that case, there is no need to push through the final mile to achieve mastery.[1]

But if you want to start *doing* physics, if you want to create something genuinely innovative, something truly new that moves the world forwards, you need to get into the habit of applying what learn.

Formulated differently, through passive learning you only create knowledge. But through active practice, you create a skill. And skill is what you need to create things on your own.

[1] As usual, the remaining "20%" that you need to acquire through active practice require around "80%" of the total effort. (This is known as the Pareto principle.)

To use an analogy, you can spend weeks researching the best instructions on the bench press technique, but you only build strength if you actively start lifting weights.

Thus, once you've read *one* textbook from cover to cover and wrote down how you would explain things to an imaginary friend, it's time to practice.

Luckily, there are many excellent textbooks on any topic that contain lots of exercises. You should, however, only use textbooks which also offer solutions to all exercises. Otherwise, you will not be able to understand what you're doing wrong. Moreover, you should never spend days trying to solve one particular exercise. Instead, whenever you struggle you should, after an hour of so, jump to the solution and try to understand it. This is far more effective since often it's just some mathematical trick you simply don't know about and there is no point in reinventing all these mathematical tricks yourself.

A key here is to recognize that physics exercises are really just a matter of practice. Once you've been exposed to lots of exercises (and their solutions) you will find it easier and easier to solve additional problems.

Next, let's talk about two additional ideas that you might find quite helpful in your journey.

23

Time

Studying physics can be tiring and frustrating at times. Thus, it is important to recognize early on that no one jumps from zero knowledge of a theory to perfect understanding in a few weeks or so. Learning physics is simply not a linear process and there is no ladder you can climb.

Instead, it's crucial to revisits the same concepts and ideas over and over again. Each time, you will view them from a different perspective and learn something new. Ideally, you should update your notes every time you learn something new.

For example, it took me roughly 6 years before I felt that I had answers to most of the questions that were bothering me about classical mechanics. Of course, I didn't spend 6 years studying classical mechanics. But I revisited various aspects of it during the years again and again. Each time I learned something new, I expanded my notes and ultimately this is how I ended up writing No-Nonsense

Classical Mechanics[1].

Don't worry that you waste time by focusing on other things for a while. In fact, quite often the only way to make progress is to ignore a topic for a while. The most profound insights usually arise during such pauses in which your brain has time to digest what you previously learned.

So don't try to rush things and give yourself time. There is no finishing line, it's not a race and all the fun really is in the learning process.

[1] Jakob Schwichtenberg. *No-Nonsense Classical Mechanics : a student-friendly introduction.* No-Nonsense Books, Karlsruhe, Germany, 2019b. ISBN 9781096195382

24

Beyond Dogma

"You do not understand an argument, until you've found the major flaws in it. For any problem complex enough to be interesting, there is evidence pointing in multiple directions." **Tyler Cowen**

"If you understand something in only one way, then you don't really understand it at all. The secret of what anything means to us depends on how we've connected it to all other things we know. Well-connected representations let you turn ideas around in your mind, to envision things from many perspectives until you find one that works for you. And that's what we mean by thinking!" **Marvin Minsky**

"Do I contradict myself?
Very well then I contradict myself,
(I am large, I contain multitudes.)" **Walt Whitman**

In can be extremely useful to have some meta understanding of how learning and understanding happens. In particular, with a concrete model at hand, you can assess your own progress and make better decisions on what to do next.

While there are many models that try to encapsulate how learning and understanding works, I recently came across one particular model that I keep thinking about and find extremely useful.

The model is a simple three-level model and was proposed by Nat Eliason.[1]

[1] https://www.nateliason.com/blog/level-3-thinking

The model describes remarkably well how I reached maturity in my thinking about different physics topics and since Nat doesn't mention physics, I want to discuss some examples below.

But first, a summary of the model:

- **Level 1** is called "**Blind Ideology**". Everyone starts at this stage for any given topic. At this stage we wholesale adopt a set of beliefs and attitudes that were thrust upon us through our upbringing or the education system. A great example is diet. Here, Level 1 means that you eat what your parents taught you to eat, which in my case was the standard Western diet.
- **Level 2** is called "**Chosen Ideology**". At this stage, people realize that the first best thing they were taught isn't the best thing that exists, and they become obsessed with another ideology. A good heuristic is that whenever someone is annoying about something, he is probably at Level 2. We reach Level 2 after a "Moment of Clarity". During such moments we realize that we have been driving with blinders on.

For the diet example above, Level 2 means that you become obsessed with something like low-carb, Paleo, veganism etc. At this stage, you are convinced that, for example, Paleo is the only way to go and every other way to eat is stupid.

- Finally, there is **Level 3**, which is called "**Ideology Transcendence**". At this stage, we can sample the best bits from prepackaged belief systems. At Level 3 we realize that no prepackaged ideology is a perfect fit for us, and we start developing our own. We start studying all ideologies that are out there and pick from each one only those parts that are of use for us.
  The step from Level 2 to Level 3 is only possible through lots of moments of clarity. Only when we are exposed to lots of contradictory points of view, we can recognize the flaws in each prepackaged belief systems. To reach Level 3 we must read books and articles that make us uncomfortable.
  Regarding the diet example, Level 3 means that you recognize that different people respond differently to different diets. Everyone has different genes and therefore, everyone has to experiment to find a diet that is a good fit. However, no prepackaged diet can be a perfect fit for everyone.
  A good test if you've already reached Level 3 is whether or not you react emotionally to new information. An emotional reaction to new information is a clear sign of Level 1 and Level 2 thinking. At Level 3 you have a well-rounded stance on the topic and primarily care about improving your understanding of it.

It's important to note that at Level 3 there is a "Strange Loop".[2] After enough time you will build an ideology of your own by picking the best stuff from other ideologies and adding something of your own. However, as soon as

[2] The notion "Strange Loop" was coined by Douglas Hofstadter in his book Gödel, Escher, Bach" [Hofstadter, 1999]: "The Strange Loop phenomenon occurs whenever, by moving upwards (or downwards) through levels of some hierarchical system, we unexpectedly find ourselves right back where we started."

this happens you are again back at Level 2 since you are again following an ideology. Then, you must again search for flaws in your thinking and get exposure to contrarian points of view. In other words, Level 3 starts again. Level 3 is a stage of constant deliberate uncertainty.

In some sense, this a miniature version of the whole scientific process. We can never know anything in the real world with 100% certainty. The only thing we can talk about is the level of confidence we have in a given theory, model or idea.

Ultimately, today's paradigm-shifting theory will become tomorrows standard theory and will again be replaced by another paradigm-shifting theory.

Nat Eliason discusses several other examples and most importantly ways to actively "level up". His essay is much better than this short summary and I highly recommend reading it.[3]

[3] As mentioned above, you can find the essay at `https://www.nateliason.com/blog/level-3-thinking`.

But now, let's discuss what all this means for physics.

24.1 Physics Beyond Dogma

Quantum Mechanics

- ▷ Level 1 is the standard "Shut up and Calculate" approach that everyone learns in the lectures and standard textbooks.
- ▷ Level 2 thinking is becoming obsessed with, for example, "Bohmian Mechanics" or the Everretian "Many-Worlds interpretation".
- ▷ Level 3 thinking is realizing that none of these approaches

is *the* answer and starting to develop your own way of thinking about quantum mechanics.

Gauge Symmetry

- Level 1 thinking is that gauge symmetry is a neat trick to derive the Lagrangian of the Standard Model and otherwise only necessary to prove renormalizability.
- Level 2 thinking is becoming obsessed with the geometrical interpretation of gauge symmetry in terms of fiber bundles or with the idea that gauge symmetries aren't fundamentally important after all, but merely redundancies in our description.
- Level 3 is when you realize that gauge symmetries are indeed only redundancies, but carry a lot of physical meaning that isn't captured by fiber bundles or the "neat idea" narrative.

General Relativity

- Level 1 is the conventional narrative that in general relativity there is no longer a gravitational field, but instead, gravity is merely a result of the curvature of spacetime.
- Level 2 is the realization that you can turn this whole idea around and argue that the essence of general relativity is that there is no spacetime at all; only interacting fields. The only thing that exists are points where spacetime trajectories of field excitations meet. This way spacetime emerges. Another possible Level 2 understanding is "GR is the unique theory with no absolute object", as coined James L. Anderson in his book Principles of Relativity Physics.[4]

[4] I have a friend who is quite obsessed with this idea.

▷ Level 3 is... I have no idea. I find the level 2 idea outlined above extremely cool and I guess this means I am stuck at level 2 for now. But if you know any articles that could help me improve beyond Level 2, please send them my way.

24.2 Leveling Up

After reading Nat Eliason's essay I started thinking about how I could use it to improve my learning process.

I started by assessing at what level I'm currently at for various topics. Unsurprisingly, it turned out I'm still at level 1 or 2 for many physics topics.

Then I started to think about how I can get from Level 1 to Level 2. The crucial step here is recognizing that there is more than what we learn in lectures and the standard textbooks.

Level 2 ideas usually can't be found in the standard textbooks. Instead, they must be actively discovered. Often it's just a side remark in a paper, book, blog post or at StackExchange that initiates the moment of clarity. Afterward, comes a period of "going down the rabbit hole" where I try to trace any reference and comment on the alternative approach.

Finally, after enough research, I slowly realize that the alternative approach I became obsessed with is not the final answer. Level 3 thinking requires that I recognize that there is more than one reasonable idea of how to go beyond what we learned in lectures and standard textbooks.

To stay at Level 3, I must be constantly exposed to ideas that challenge my current beliefs. If I become too certain of a given idea I fall back to Level 2. Level 3 is uncomfortable and lonely.

To summarize: To level up you must read broadly. If you only stick to the books that your professor recommends, you will stay at Level 1. Read books and articles by experts, read blog posts, read comments at StackExchange or at the PhysicsForums, read stuff by weird unknown guys. It doesn't matter as long as they do not all repeat the standard story over and over again. As soon as some alternative approach sparks your interest, it is necessary to dig deep and understand it from all possible angles. While it is extremely helpful to become obsessed during this phase, this obsession should always end after some time. At some point, it is necessary to recognize that there is no universal prepackaged answer.

The most important message to take away from all of this is probably that you always need to stay humble, keep studying, and trust the process. At some point, you'll reach a stage at which you're not only confident in your understanding but at which your confidence is actually warranted.

25

Closing Words

When you learn physics from textbooks, it can often seem as if physics were almost a finished project. But this couldn't be further from the truth.

There are still *lots* of open questions.

Most importantly, there is still no generally accepted way to incorporate general relativity in the framework of quantum field theory. While Einstein's classical field model works perfectly at large scales, we know that at small scales quantum field theory holds sway.[1] But there are lots of other more modest open questions, for example, about the role of torsion in general relativity.

[1] If you're interested in quantum gravity, you should start by reading Three Roads to Quantum Gravity by Lee Smolin [Smolin, 2017]. It's non-technical and unlike most other books on the subject undogmatic.

Moreover, there are lots of open questions about the particle zoo. Why exactly these particles? Why do they carry the charges they do? In addition, there are good reasons to believe that there are additional elementary particles, for example, to explain the masses of neutrinos and the observed dark matter density.

I think there is still much to come and maybe a completely new framework is needed to overcome the present obstacles. Future developments will be fascinating and I hope you will continue to follow the story and maybe contribute something yourself.

One Last Thing

It's impossible to overstate how important reviews are for an author. Most book sales, at least for books without a marketing budget, come from people who find books through the recommendations on Amazon. Your review helps Amazon figure out what types of people would like my book and makes sure it's shown in the recommended products.

I'd never ask anyone to rate my book higher than they think it deserves, but if you like my book, please take the time to write a short review and rate it on Amazon. This is the biggest thing you can do to support me as a writer.

Each review has an impact on how many people will read my book and, of course, I'm always happy to learn about what people think about my writing.

PS: If you write a review, I would appreciate a short email with a link to it or a screenshot to Jakobschwich@gmail.com. This helps me to take note of new reviews. And, of course, feel free to add any comments or feedback that you don't want to share publicly.

Part VI
Appendix

A

FAQ

I update this FAQ regularly. So if you have any kind of question, email me at jakobschwich@gmail.com

I'm stuck. What should I do?

In short: ask. You need to be brave enough to make a fool of yourself. Many of the questions you will ask will turn out to be stupid. But to quote Confucius:

> "The man who asks a question is a fool for a minute, the man who does not ask is a fool for life."

Of course, you will feel stupid all the time while everyone else acts smart. But ultimately, you will be the one who truly understands things while everyone else remains on the same superficial level. Unfortunately, most students and even professors are too afraid to ask fearing "sounding dumb".

So, whenever you feel stuck, try to identify what exactly confuses you. Oftentimes once you're able to formulate your problem in terms of a concrete question, much of the confusion and frustration vanishes immediately. If not, enter your question in the search engine of your choice (Google, DuckDuckGo etc.) and try to find an answer. Pay special attention to search results from Physics.Stackexchange.com and similar QnA platforms.

If you can't find a good answer, ask your question at `www.Physics.Stackexchange.com`, `www.PhysicsOverflow.org` and `www.Physicsforums.com`. Never be afraid to ask questions, especially if you fear that it may be too trivial. Too many "trivial" questions never get asked and keep on bothering students. The ability to ask "basic question" is a superpower. In addition, especially in physics, many "trivial" questions often lead to extremely deep insights.

If you get a mean reply or comment, it simply means that the commentator is having a bad day and that's certainly nothing you should care about.[1] Good questions and the resulting answers often help hundreds of other students all around the world.

[1] Be warned that some people have *lots* of bad days.

A great advantage of asking your questions online is that you can post them anonymously. This helps to avoid the fear that others think you're stupid because you ask such trivial questions.[2] In addition, of course, you can get an answer from anyone worldwide, and they are saved such that you and others can revisit them in the future.

[2] This is the main reason why it's so rare that anyone asks questions in an in-person lecture.

Sometimes it takes a few days until you get a good answer. In the meantime, simply move on and try to focus on something else. It usually makes no sense to dwell for days or even weeks on a specific problem or question because the chances are extremely high that someone more experienced can help you solve it in a few minutes. Asking your questions on a platform like StackExchange is a good way to get them out of your head. You can always come back and revisit all your questions in your profile. Maybe, after some time you will be able to answer them yourself. If this is the case, post your answer online. This helps to avoid that you'll get confused by the same thing again and potentially helps lots of other students.

What about video lectures?

Video lectures, like in-person lectures, have the tendency to always move at the wrong pace. Lecturers often dwell endlessly on things you already know and move far too quickly past complicated concepts.

In contrast, a book allows you to move at your own pace. You can reread paragraphs you didn't fully understand and read them as slowly as you wish. In addition, a book allows you to jump around within the book and between books. You can easily return to any paragraph whenever you need a refresher. This is essential because, as mentioned in Chapter 23, physics cannot be learned linearly. Only by revisiting concepts and ideas over and over again from various perspectives, you'll develop a deep understanding. While this is, theoretically, also possible with video lectures it's a lot more cumbersome.

In addition, if you're watching a video lecture, distractions like Reddit and Twitter are always only a click away. A book allows you to get away from your laptop. Ideally, books are read at the library far away from all distractions. This is essential because you need periods of deep work to grasp difficult concepts. This is, of course, only my opinion and there may be people who are able to learn physics from video lectures. However, all students I know who focused on video lectures never got very far. This observation is also supported by the extremely low completion rates of MOOCs[3]. In addition, it's simply a myth that there are different 'Learning Styles'[4].

[3] https://www.newyorker.com/science/maria-konnikova/moocs-failure-solutions

[4] https://www.theguardian.com/education/2017/mar/12/no-evidence-to-back-idea-of-learning-styles https://www.theatlantic.com/science/archive/2018/04/the-myth-of-learning-styles/557687/

What about teachers and professors?

As discussed already in Chapter 21, no teacher will be able to answer all your questions. But nevertheless it's true that a great teacher or tutor (or coach to use a more modern term) can be immensely helpful.

In particular, the best teachers will gently nudge you in the right direction and then let you figure it out yourself. If you ask for a fish, they will hand you a worm. If you ask a question, they will answer by raising a new question or referencing a resource that allows you to understand the question from a new perspective. This is a far more effective approach because, as mentioned already in Chapter 21, the only path you know by heart is the one you discovered yourself. You might be disappointed at first, but that's the only way it's going to work.

This is really how you can distinguish between a great teacher and a charlatan. A charlatan will act as if he or she has all the answers and will get annoyed quickly if you ask follow-up questions. But a great teacher who truly wants you to succeed, freely admits holes in his own understanding and supports you in discovering things on your own terms.

What about the standard books?

We talked already about this in Chapter 20. Nevertheless, since this is so important and one of the most common pitfalls, let me repeat my arguments in slightly different words.

The standard books are commonly used and recommended in lectures because professors love them. But a good textbook is, first and foremost, loved by students. And usually students and professors do not like the same books because most professors have forgotten what it's like to be a beginner. Professors like to recommend books which are very complete, concise and exact, while students need books which tell them (unavoidably sometimes in vague and verbose terms) what is really going on and explain things using analogies.

Moreover, you need to be aware that signalling always also plays a role when it comes to book recommendations.

If you recommended a complicated and hard-to-understand book, you signal that you're smart. If you recommend "Quantum Mechanics for Dummies" you signal, well, that you're a person who needs a dummy explanation to understand the topic. This is a big factor why many awful books get recommended over and over again.

In addition, it's much safer to recommend a standard book. If you recommend a book that thousands of others recommend, no one will blame you. But if you recommend a book only a few people know of, it's quite easy to criticize you for your recommendation.

Nevertheless, all standard books are mentioned in the subject guides because sometimes they are useful as a reference once you've understood the subject.

Bibliography

Luis Alvarez-Gaumé and Miguel Vázquez-Mozo. *An invitation to quantum field theory.* Springer, Heidelberg New York, 2012. ISBN 978-3-642-23728-7.

David Appell. When supergravity was born. *Physics World*, 25(09):32–36, sep 2012. DOI: 10.1088/2058-7058/25/09/36.

Dan Ariely. *Predictably irrational : the hidden forces that shape our decisions.* Harper Perennial, New York, 2010. ISBN 9780061353246.

V. I. Arnold. *Mathematical methods of classical mechanics.* Springer-Verlag, New York, 1989. ISBN 9780387968902.

V. I. Arnold. Opinion. *The Mathematical Intelligencer*, 17(3): 6–10, Jan 1995. ISSN 0343-6993. DOI: 10.1007/BF03024363. URL `https://doi.org/10.1007/BF03024363`.

Isaac Asimov. *Science past, science future.* Doubleday, Garden City, N.Y, 1975. ISBN 9780385099233.

Robert Batterman. *The devil in the details.* Oxford University Press, Oxford New York, 2002. ISBN 9780195314885.

Bo Bennett. *Logically fallacious : the ultimate collection of over 300 logical fallacies.* Archieboy Holdings, LLC, Sudbury, MA, 2015. ISBN 978-1456607524.

Peter Bevelin. *Seeking wisdom : from Darwin to Munger*. PCA Publications, Malmoe, Sweden, 2013. ISBN 978-1578644285.

Jorge Borges. *Professor Borges : a course on English literature*. New Directions Books, New York, 2013. ISBN 9780811218757.

Tian Cao. *Conceptual foundations of quantum field theory*. Cambridge University Press, New York, 1999. ISBN 9780511470813.

Sean Carroll. *Spacetime and geometry : an introduction to general relativity*. Pearson, Harlow, Essex, 2014. ISBN 9781292026633.

T.P. Cheng. *Relativity, gravitation and cosmology : a basic introduction*. Oxford University Press, Oxford New York, 2010. ISBN 9780199573646.

F. E. Close. *The infinity puzzle : quantum field theory and the hunt for an orderly universe*. Basic Books, New York, 2011. ISBN 9780465063826.

Claude Cohen-Tannoudji. *Quantum mechanics*. Wiley, New York, 1977. ISBN 9780471164333.

Jennifer Coopersmith. *The lazy universe : an introduction to the principle of least action*. Oxford University Press, Oxford New York, NY, 2017. ISBN 978-0-19-874304-0).

P. C. W. Davies. *The New physics*. Cambridge University Press, Cambridge Cambridgeshire New York, 1989. ISBN 9780521304207.

Emanuel Derman. *My life as a quant : reflections on physics and finance*. Wiley, Hoboken, N.J, 2004. ISBN 9780470192733.

Carol Dweck. *Mindset : the new psychology of success*. Random House, New York, 2006. ISBN 9780345472328.

Anders Ericsson and Robert Pool. *Peak : secrets from the new science of expertise*. Houghton Mifflin Harcourt, Boston, 2016. ISBN 978-0544456235.

Howard Eves. *Mathematical circles*. Mathematical Association of America, Washington, D.C, 2003. ISBN 9780883855430.

Lizhi Fang and Remo Ruffini. *Quantum cosmology*. World Scientific, Singapore Teaneck, NJ, USA, 1987. ISBN 978-9971502935.

Graham Farmelo. *The strangest man : the hidden life of Paul Dirac, quantum genius*. Faber and Faber, London, 2009. ISBN 9780571222865.

Peter Feibelman. *A PhD is Not Enough! : A Guide To Survival in Science*. Basic Books, New York, 2011. ISBN 9780465022229.

Paul Feyerabend. *Against method*. Verso, London New York, 2010. ISBN 978-1844674428.

Richard Feynman. *Surely You're Joking, Mr. Feynman!" : Adventures of a Curious Character*. W.W. Norton, New York, 1985. ISBN 978-0393316049.

Daniel Fleisch. *A student's guide to Maxwell's equations*. Cambridge University Press, Cambridge, UK New York, 2008. ISBN 978-0-511-38900-9.

Matthew Frederick. *101 things I learned in architecture school*. MIT Press, Cambridge, Mass, 2007. ISBN 978-0262062664.

Steven French. *Philosophy of science : key concepts*. Bloomsbury Academic, London New York, 2016. ISBN 9781474245234.

Max Frisch. *Gantenbein : a novel*. Harcourt Brace Jovanovich, San Diego, 1982. ISBN 978-0156344074.

G. Gamow, D. Ivanenko, and L. Landau. World constants and limiting transition. *Physics of Atomic Nuclei*, 65(7): 1373–1375, Jul 2002. ISSN 1562-692X.

Gian Francesco Giudice. The Dawn of the Post-Naturalness Era. In Aharon Levy, Stefano Forte, and Giovanni Ridolfi, editors, *From My Vast Repertoire ...: Guido Altarelli's Legacy*, pages 267–292. 2019.

Malcolm Gladwell. *Outliers : The Story of Success*. Little, Brown and Company, New York, 2008. ISBN 978-0316017930.

Peter Godfrey-Smith. *Theory and reality : an introduction to the philosophy of science*. University of Chicago Press, Chicago, 2003. ISBN 978-0226300634.

John Goold, Marcus Huber, Arnau Riera, Lídia del Rio, and Paul Skrzypczyk. The role of quantum information in thermodynamics—a topical review. *Journal of Physics A: Mathematical and Theoretical*, 49(14):143001, feb 2016. DOI: 10.1088/1751-8113/49/14/143001.

David Griffiths. *Introduction to quantum mechanics*. Pearson Prentice Hall, Upper Saddle River, NJ, 2005. ISBN 9780131118928.

David Griffiths. *Introduction to elementary particles*. Wiley-VCH, Weinheim Germany, 2008. ISBN 9783527406012.

David Griffiths. *Introduction to electrodynamics*. Cambridge University Press, Cambridge, United Kingdom New York, NY, 2017. ISBN 9781108420419.

J. B. Hartle. *Gravity : an introduction to Einstein's general relativity*. Addison-Wesley, San Francisco, 2003. ISBN 9780805386622.

Banesh Hoffmann. *Albert Einstein, creator and rebel*. Viking Press, New York, 1972. ISBN 978-0670111817.

Douglas Hofstadter. *Gödel, Escher, Bach : an eternal golden braid*. Basic Books, New York, 1999. ISBN 9780465026562.

Sabine Hossenfelder. *Lost in math : how beauty leads physics astray*. Basic Books, New York, NY, 2018. ISBN 9780465094257.

Andrew Hunt. *Pragmatic thinking and learning : refactor your "wetware*. Pragmatic, Raleigh, 2008. ISBN 978-1934356050.

John Jackson. *Classical electrodynamics*. Wiley, New York, 1999. ISBN 9780471309321.

E. T. Jaynes. Gibbs vs Boltzmann Entropies. *American Journal of Physics*, 33:391–398, May 1965. DOI: 10.1119/1.1971557.

E. T. Jaynes. The second law as physical fact and as human inference, 1998.

Daniel Kahneman. *Thinking, fast and slow*. Farrar, Straus and Giroux, New York, 2011. ISBN 9780141033570.

Karen Kelsky. *The professor is in : the essential guide to turning your Ph.D. into a job*. Three Rivers Press, New York, 2015. ISBN 978-0553419429.

Boris Khesin. *Arnold : swimming against the tide*. American Mathematical Society, Providence, Rhode Island, 2014. ISBN 9781470416997.

Robert Klauber. *Student friendly quantum field theory : basic principles and quantum electrodynamics*. Sandtrove Press, Fairfield, Iowa, 2014. ISBN 9780984513956.

Martin Krieger. *Doing physics : how physicists take hold of the world*. Indiana University Press, Bloomington, 1992. ISBN 978-0253207012.

Thomas Kuhn. *The structure of scientific revolutions.* The University of Chicago Press, Chicago London, 2012. ISBN 9780226458113.

Cornelius Lanczos. *The variational principles of mechanics.* Dover Publications, New York, 1986. ISBN 9780486650678.

Ian Lawrie. *A unified grand tour of theoretical physics.* CRC Press, Taylor & Francis Group, Boca Raton, 2013. ISBN 978-1439884461.

Jean-Marc Lévy-Leblond. One more derivation of the Lorentz transformation. *American Journal of Physics*, 44(3): 271–277, Mar 1976. DOI: 10.1119/1.10490.

Michael Lewis. *The undoing project : a friendship that changed our minds.* W.W. Norton & Company, New York, 2017. ISBN 9780241254738.

Malcolm Longair. *Theoretical concepts in physics : an alternative view of theoretical reasoning in physics.* Cambridge University Press, Cambridge, U.K. New York, 2003. ISBN 978-0521528788.

Joao Magueijo. *Faster than the speed of light : the story of a scientific speculation.* Arrow Books, London, 2004. ISBN 978-0099428084.

Sanjoy Mahajan and Richard R. Hake. Is it time for a science counterpart of the Benezet-Berman mathematics teaching experiment of the 1930's? *arXiv e-prints*, art. physics/0512202, Dec 2005.

N. David Mermin. Making better sense of quantum mechanics. *Reports on Progress in Physics*, 82(1):012002, Jan 2019. DOI: 10.1088/1361-6633/aae2c6.

Haruki Murakami. *Norwegian wood.* Vintage, London, 2003. ISBN 9780099448822.

Roland Omnes. *The interpretation of quantum mechanics*. Princeton University Press, Princeton, N.J, 1994. ISBN 978-0691036694.

Roger Penrose. *The road to reality : a complete guide to the laws of the universe*. A.A. Knopf, New York, 2005. ISBN 978-0679454434.

Matthew Robinson. *Symmetry and the standard model : mathematics and particle physics*. Springer, New York, 2011. ISBN 9781441982667.

Cosma Rohilla Shalizi. The Backwards Arrow of Time of the Coherently Bayesian Statistical Mechanic. *arXiv e-prints*, art. cond-mat/0410063, Oct 2004.

Laura Ruetsche. *Interpreting quantum theories : the art of the possible*. Oxford University Press, Oxford, 2013. ISBN 978-0199681068.

Miguel Ruiz. *The Four Agreements : a Practical Guide to Personal Freedom*. Amber-Allen Publishing, San Rafael, California Carlsbad, California, 1997. ISBN 9781878424310.

Jeff Schmidt. *Disciplined Minds*. Rowman & Littlefield, Lanham, Md, 2000. ISBN 9780742516854.

Bernard Schutz. *A first course in general relativity*. Cambridge University Press, Cambridge New York, 2009. ISBN 9780521887052.

Matthew Schwartz. *Quantum field theory and the standard model*. Cambridge University Press, Cambridge, United Kingdom New York, 2014. ISBN 9781107034730.

Jakob Schwichtenberg. *Physics from symmetry*. Springer, Cham, Switzerland, 2018a. ISBN 978-3-319-66631-0.

Jakob Schwichtenberg. *No-nonsense quantum mechanics : a student-friendly introduction*. No-Nonsense Books, Karlsruhe, Germany, 2018b. ISBN 9781790455386.

Jakob Schwichtenberg. Demystifying Gauge Symmetry. 2019a.

Jakob Schwichtenberg. *No-Nonsense Classical Mechanics : a student-friendly introduction*. No-Nonsense Books, Karlsruhe, Germany, 2019b. ISBN 9781096195382.

Jakob Schwichtenberg. *No-Nonsense Electrodynamics*. No-Nonsense Books, Karlsruhe, Germany, 2019c. ISBN 9781790842117.

Jakob Schwichtenberg. *Physics from finance*. No-Nonsense Books, Karlsruhe, Germany, 2019d. ISBN 978-1795882415.

Ramamurti Shankar. *Principles of quantum mechanics*. Plenum Press, New York, 1994. ISBN 9780306447907.

Kevin Simler and Robin Hanson. *The elephant in the brain : hidden motives in everyday life*. Oxford University Press, New York, NY, 2018. ISBN 9780190495992.

Karoly Simonyi. *A cultural history of physics*. CRC Press, Boca Raton, Fla, 2012. ISBN 9781568813295.

Lawrence Sklar. *Physics and chance : philosophical issues in the foundations of statistical mechanics*. Cambridge University Press, Cambridge England New York, 1993. ISBN 978-0521558815.

Lee Smolin. Why no 'new einstein'? *Physics Today*, 6(58):56, 2005.

Lee Smolin. *The trouble with physics : the rise of string theory, the fall of a science and what comes next*. Penguin, London, 2008. ISBN 9780141018355.

Lee Smolin. *Three roads to quantum gravity*. Basic Books, New York, 2017. ISBN 9780465094547.

J. Stachel. A Brief History of Space-Time. In I. Ciufolini, D. Dominici, and L. Lusanna, editors, *2001: A Relativistic Spacetime Odyssey*, pages 15–34, January 2003.

Andrew M. Steane. An introduction to spinors, 2013.

Kurt Sundermeyer. *Symmetries in fundamental physics.* Springer, Cham, 2014. ISBN 9783319065809.

Nassim Taleb. *Antifragile : things that gain from disorder.* Random House, New York, 2012. ISBN 9781400067824.

Edwin Taylor and John Wheeler. *Spacetime physics : introduction to special relativity.* W.H. Freeman, New York, 1992. ISBN 9780716723271.

William P. Thurston. On proof and progress in mathematics, 1994.

Tommaso Toffoli. Entropy? honest! *Entropy*, 18(7), 2016. ISSN 1099-4300. DOI: 10.3390/e18070247. URL `https://www.mdpi.com/1099-4300/18/7/247`.

Peter Woit. *Not even wrong : the failure of string theory and the continuing challenge to unify the laws of physics.* Jonathan Cape, London, 2006. ISBN 9780224076050.

Scott Young. *Ultralearning: seven strategies for mastering hard skills and getting ahead.* Harper Collins Publ., UK, 2019. ISBN 9780008305703.

Andrew Zangwill. *Modern electrodynamics.* Cambridge University Press, Cambridge, 2013. ISBN 978-0-521-89697-9.

A Zee. *Quantum field theory in a nutshell.* Princeton University Press, Princeton, N.J, 2010. ISBN 9780691140346.

A Zee. *Einstein gravity in a nutshell.* Princeton University Press, Princeton, 2013. ISBN 9780691145587.

Made in the USA
Columbia, SC
18 July 2020